QUATERNARY OF SOUTH AMERICA
AND ANTARCTIC PENINSULA

VOLUME 5

LESTODON ARMATUS Gervais

QUATERNARY OF SOUTH AMERICA AND ANTARCTIC PENINSULA

*With selected papers of the special session on
the Quaternary of South America
XIIth INQUA International Congress
Ottawa, 31 July-9 August 1987*

Edited by
JORGE RABASSA
*Centro Austral de Investigaciones Científicas, Ushuaia
Tierra del Fuego*

VOLUME 5 (1987)

A.A.BALKEMA / ROTTERDAM / BROOKFIELD / 1987

The texts of the various papers in this volume were set
individually by typists under the supervision of each of
the authors concerned.

ISSN 0168-6305

ISBN 90 6191 733 6

For USA & Canada: A.A.Balkema Publishers, Old Post Road,
Brookfield, VT 05036, USA

Printed in the Netherlands

Contents

SPECIAL SESSION ON THE QUATERNARY OF SOUTH AMERICA, XIIth INQUA
INTERNATIONAL CONGRESS

Obituary
Daniel Alberto Valencio (1928-1987)

Daniel Alberto Valencio

Daniel A. Valencio, a pioneer of research on Paleomagnetism
and its geodynamic implications, died at Buenos Aires on May
28th, 1987, a few weeks before his 59th birthday. Geophysicists
all over the world, and especially his pupils in Bolivia,
Brazil, Great Britain and Mexico will join their Argentine
colleagues in deploring his ultimately passing away.

A graduate from La Plata University, Valencio started his
professional career at the Argentine National Oil Fields
Administration as an Exploration Geophysicist in 1952; from
1960 to 1962, he held the position of Head of the Geophysical
Department at the Cuban Institute of Oil and Mineral Resources.
After his return to Argentina, he became Professor of
Geophysics at Buenos Aires University. During the 25 years of
his association with the Department of Geological Sciences
there, he succeeded in establishing an admirably close
interdisciplinary connection between the geological and
geophysical aspects of solid earth problems.

In 1964 Valencio laid the foundations of the Laboratory of
Paleomagnetism which was to become the centre of an impressive
and ever-increasing activity in research and teaching, including
several international post-graduate courses. Numerous students
of Geology and Physics sought his guidance for their under-
graduate and graduate studies, and many of them remained
associated with his group as investigators. In addition to
his unswerving teaching activities at Buenos Aires, Valencio
was visiting Professor in Brazil and Mexico. He was an active,
and in some cases, leading participant, both on a national
and world-wide scale, in the planning, coordination and
execution of several great interdisciplinary projects such
as the Upper Mantle Project, the International Geodynamics
Project and the International Lithosphere Project, becoming
the chairman or cochairman of diverse commitees and working
groups as well as organizer of several symposia. He was a
member of the Executive Committee of the Inter-Union Commission
on Geodynamics (1970-1980) and leader of its Working Group on
"Global synthesis of evidence leading to the reconstruction
of the distribution of continents and oceans through time."

In the framework of IAGA, Valencio served as a member of its
Executive Committee during the 1979-1983 term, as chairman
of the Working Group on Paleomagnetism in the following four
years period, and also as liaison officer with other bodies
of IUGG. At home, he was a top-ranking Research Fellow of
the Consejo Nacional de Investigaciones Científicas y Técnicas;
member of diverse advisory committees of the "Consejo";
chairman of the National Committee for the Inter-Union
Lithosphere Project, and president of the Asociación Argentina
de Geofísicos y Geodestas (AAGG) during one four years term.
He was the author of a book on "El magnetismo de las rocas"
(1980), and co-editor of three other books dealing with the
interdisciplinary aspects of Paleomagnetism, as well as member
of several editorial boards, Valencio was an Honorary Fellow
of the AAGG, an Ordinary Member of the Academia Nacional de
Geografía, and a Foreign Member of the Royal Astronomical
Society, whose "Geophysical Journal" published several of his
papers.

In pondering Valencio's overwhelming productivity in the
fields of teaching, research, academic administration, and
coordination of interdisciplinary projects, both local and
international, one cannot help suspecting that some kind of
foreboding may have challenged him to generously give his

2

utmost during the limited span that has fallen to his lot.
   Daniel is survived by Nike, his wife, three daughters
(including a geologist), and a son, also a student of Earth
Sciences.

Otto Schneider
Universidad de Buenos Aires - CONICET

This obituary has been previously published by the
International Association of Geomagnetism and Aeronomy, June
6th., 1987, and it is reproduced with permission of the author.

DANIEL A. VALENCIO† & M.L. BOBBIO
Departamento de Ciencias Geológicas, Ciudad Universitaria, Buenos Aires, Argentina

M.J. ORGEIRA
Consejo Nacional de Investigaciones Científicas y Técnicas, Argentina

2

Magnetostratigraphy and magnetic susceptibility data of the late Cenozoic 'Ensenadense' and 'Bonaerense' sediments of Buenos Aires and La Plata, Argentina

ABSTRACT

Paleomagnetic and magnetic susceptibility data for sequences of late Cenozoic "Ensenadense" and "Bonaerense" sediments exposed by excavations carried out in the Cities of Buenos Aires (lat. 34.5° S; long. 58.5° W) and La Plata (lat. 35° S; long. 58° W) are reported. These geological units are carriers of land mammal fauna assemblages younger than the Great American Faunal Interchange.

The lithology as well as the paleomagnetic and magnetic susceptibility data indicate that the "Ensenadense" and "Bonaerense" sediments from these cities were accumulated under different geological processes.

In Buenos Aires, the accumulation of these late Cenozoic sediments was practically continuous. Only one period of no-accumulation and/or erosion was recorded in the "Ensenadense" sediments at about 13.15 m below ground level. The magneto-stratigraphy for this sedimentary sequence suggests a predominantly Brunhes (<0.7 Ma, middle-late Pleistocene) magnetic age for the "Bonaerense" sediments and a Matuyama to late Gauss (>0.7 Ma- <2.84 Ma, late Pliocene-middle Pleistocene) magnetic age for the "Ensenadense" deposits.

In La Plata, the accumulation of the "Ensenadense" and "Bonaerense" sediments was probably discontinuous. The magnetostratigraphy for these sediments suggests two interpretations, which would indicate middle-late Brunhes magnetic age for the "Ensenadense" and "Bonaerense" sediments from the uppermost 14.0 m of the sequence. One of the interpretations suggests a late Gauss-Matuyama magnetic age (<3.0 - >0.7 Ma; late Pliocene-middle Pleistocene) for the

"Ensenadense" sediments from the lowest 14.0 m of the sequence.
The other interpretation would indicate, instead, a Matuyama
($<$2.41 Ma - $>$0.7 Ma, late Pliocene-middle Pleistocene)
magnetic age for these "Ensenadense" sediments.

RESUMEN

Se presentan en este trabajo datos paleomagnéticos y de sus-
ceptibilidad magnética obtenidos para secuencias de sedimentos
"Ensenadenses" y "Bonaerenses" del Cenozoico Superior, expues-
tos en excavaciones en las ciudades de Buenos Aires (34°5 lat.S;
58°5 long. W) y La Plata (35° lat. S; 58° long. W). Estas
unidades geológicas son portadoras de asociaciones faunísticas
de mamíferos terrestres, más jóvenes que el Gran Intercambio
Faunístico Americano.

La litología, así como los datos paleomagnéticos y de sus-
ceptibilidad magnética, indican que los sedimentos "Ensenadenses" y "Bonaerenses" de estas ciudades fueron acumulados bajo
procesos geológicos diferentes.

En Buenos Aires, la acumulación de estos sedimentos fue
prácticamente continua. Solamente un período de no-acumulación
y/o erosión ha sido detectado en los sedimentos "Ensenadenses",
a aproximadamente 13,15 m bajo el nivel del suelo. La magnetoestratigrafía para esta secuencia sedimentaria sugiere una
edad magnética Brunhes predominante ($<$0,7 Ma; Pleistoceno
medio a superior) para los sedimentos "Bonaerenses" y una edad
magnética Matuyama a Gauss tardía ($>$0,7 Ma -$<$2,84 Ma, Plioceno tardío a Pleistoceno medio) para los depósitos "Ensenadenses".

En La Plata, la acumulación de sedimentos "Ensenadenses" y
"Bonaerenses" fue probablemente discontinua. La magnetoestratigrafía para estos sedimentos sugiere dos interpretaciones,
las cuales indicarían una edad magnética Brunhes media a tardía para los sedimentos "Ensenadenses" y "Bonaerenses" de los
14,0 m superiores de la secuencia. Una de las interpretaciones
sugiere una edad magnética Gauss tardía-Matuyama ($<$3,0 Ma -
$>$0,7 Ma; Plioceno tardío - Pleistoceno medio) para los sedimentos "Ensenadenses" de los 14,0 m inferiores de la secuencia.
La otra interpretación indicaría, en cambio, una edad magnética Matuyama ($<$2,41 Ma - $>$0,7 Ma; Plioceno tardío - Pleistoceno medio) para estos sedimentos "Ensenadenses".

INTRODUCTION

Throughout most of the Tertiary, South America was an isolated
continent. During most of this period, South America had a
highly distinctive fauna where unique land mammal faunas
developed. It is widely accepted that the isolation of South
America ended about 3 million years ago (late Pliocene), when
the Panamá Isthmus came into existence, uniting North and
South America. This permitted the reciprocal interchange of
terrestrial biota between both Americas. This biotic event
is known as the Great American Faunal Interchange (Webb, 1976).

Palaeontologist used this evidence to define the age of the
New World's mammals; particularly, the age of the South
American mammals in pre-(mostly endemic fauna) and post-land
bridge times (endemic plus North American fauna). The
chronological order of the South American Land Mammal ages are:
"Huayqueriense"; "Montehermosense"; "Uquiense"; "Ensenadense"
and "Lujanense" (Pascual et al., 1965, Table 1). Pascual and
Fidalgo (1972) as well as Marshall and Pascual (1978) suggest
different geological ages for these land mammal time-units
(Table 1). The beginning of the major faunal interchange
occurred during the "Montehermosense" mammal age when North-
American fossils (Cricetidae and Tayassuidae) occurred in
South America (Pascual et al., 1965). This should be related
with the development of the stable corridor between both
Americas. Hallam (1972) suggests that this occurred in the
late Pliocene, about 3 Ma; Tarling (1981) reports that "the
development of such a bridge was critical during the final
linkage between North and South America during Miocene-Pliocene
times".

Sediments exposed or circumstantially exposed in excavations
for building purposes in Buenos Aires Province, have produced
one of the richest records of vertebrate life during the Great
American Faunal Interchange. These sediments are assigned to
geological formations or units: Chapadmalal, Barranca de Los
Lobos, Vorohué, San Andrés and Miramar Formations, "Ensenadense"; Arroyo Seco and Arroyo Lobería Formations, "Bonaerense".
The ages assigned to these formations and units are summarized
in Table 1. These formations and units' relationships as well
as the South American Land Mammal Ages are also shown in this
table. It is not possible to give a more precise age to these
sediments due to the lack of precise biostratigraphic and
absolute age controls. Therefore, we have programmed a

7

Table 1

Formation or Stratigraphic Unit	Age	South American Land Mammal Ages (Pascual et al., 1965)	Geological Ages	
			(Pascual and Fidalgo, 1972)	(Marshall and Pascual, 1978)
"Bonaerense" Arroyo Lobería Fm. Arroyo Seco Fm.	middle-late Pleistocene (Frenguelli, 1957); early Pleistocene (Ameghino, 1908)	Lujanense	late Pleistocene	late Pleistocene
"Ensenadense" Miramar Fm.	middle-Pleistocene (Frenguelli, 1957); Pliocene, (Ameghino, 1908)	Ensenadense	middle Pleistocene	late-middle Pleistocene
San Andrés Fm. Vorohué Fm. Barranca de Los Lobos Fm.	Pleistocene (Kraglievich, 1952)	Uquiense	early Pleistocene	middle Pleistocene-late Pliocene

Table 1. Cont.

Formation or Stratigraphic Unit	Age	South American Land Mammal Ages (Pascual et al., 1965)	Geological Ages	
			(Pascual and Fidalgo, 1972)	(Marshall and Pascual, 1978)
Chapadmalal Fm. (equivalent to the "Hermo-sense" and "Chapadmalense")	late Miocene (Ameghino, 1908); late Pliocene (Kraglievich, 1952); early Pleistocene (Frenguelli, 1957)			
		Montehermosense	late Pliocene	late-early Pliocene
		Huayqueriense	middle Pliocene	early Pliocene- late Miocene.

systematic paleomagnetic study of sequences of late Cenozoic
sediments of Buenos Aires Province. Our purpose is to use the
magnetostratigraphy of these sequences to define the absolute
age of the sediments. Magnetostratigraphy relies on some basic
assumptions: 1) the Earth's magnetic field has reversed its
polarity in the past; 2) the timing of these reversals is
accurately known; 3) sediments record the direction of the
geomagnetic field at or near the time of deposition (primary
remanent magnetization, PRM) and 4) this PRM is preserved
throughout geological time. In many cases, it is not easy
to establish the correlation tie-lines between the magnetostratigraphy for a given sequence and the reversal time scale
for the late Cenozoic. This is also valid when datable rocks
are available in the sequence, because the uncertainty of the
radiometric age is frequently higher than the time-span of an
event of polarity of the geomagnetic field. In order to solve
that problem, the sequences of "Ensenadense" and "Bonaerense"
sediments are being studied. These are among the youngest
geological units exposed in Northeastern Buenos Aires Province
(Valencio and Orgeira, 1983; Bobbio **et al.**, 1985). Then, a
key reference mark, the Brunhes-Matuyama boundary, may be used
to establish the correlation. Paleomagnetic data for sequences
of the "Ensenadense" and "Bonaerense" sediments exposed in
large excavations in the Cities of Buenos Aires (34.5° S, 58.5°
W) and La Plata (35° S, 58° W) for building foundation are
presented here (Figure 1).

Marshall **et al** (1979, 1982), McFadden **et al** (1983) and
Butler **et al** (1984) have reported paleomagnetic data for late
Cenozoic sediments exposed in northern Argentina and Bolivia.

GEOLOGICAL SETTING AND SAMPLING SITES

Late Cenozoic sediments, bearing characteristic fossil mammal
assemblages, are exposed in different sites of Argentina. The
lithology of these sediments is remarkably homogeneous in
Buenos Aires Province (Figure 1). On the basis of lithology
and stratigraphic relationships, these sediments have been
included in formations or units (Table 1); however, it is not
always easy to define accurately the transitions between these
formations or units. Characteristic fossil mammal assemblages
have been found in these sediments. Paleontologists used them
to define land mammal ages for these sediments (Table 1).

10

Late Cenozoic sediments are poorly exposed in the Cities of
Buenos Aires and La Plata. However, sequences of these
sediments are uncovered by excavations. The "Ensenadense" was
defined by Ameghino (1889) at the City of Ensenada, 7.5 km NE
of La Plata (Figure 1). The "Ensenadense" and the underlying
"Pre-ensenadense" are, according to Ameghino, the oldest units
of the "Pampeano". This lays upon the "Puelchense" sands;
Ameghino suggested a Miocene age for the latter. The
classification used in this paper is that of Frenguelli (1957)
who included the upper part of the "Pre-ensenadense" of
Ameghino within the "Ensenadense". Frenguelli (1957) presented
a detailed description of the "Ensenadense" sediments; it is
roughly constituted by a thick brown unit of loessic silts
with calcareous nodules of different forms and an intercalation
of greenish lacustrine sediments. It generally presents
irregular stratification at its base which disappears toward
the top. Ameghino (1908) assigned the "Ensenadense" to the
Pliocene, whereas Frenguelli (1957) suggested a middle
Pleistocene age for this unit. Pascual et al (1965), Pascual
and Fidalgo (1972) and Marshall and Pascual (1978) correlated
the "Ensenadense" and "Bonaerense" with the middle Pleistocene
and middle-late Pleistocene, respectively, on basis of its
fossil mammal fauna.

Frenguelli (1957) described the "Bonaerense" as a thick,
homogeneous, fine-grained, light reddish brown unstratified
unit, with small holes left by roots and homogeneous
distribution of calcareous nodules. There is general agreement
to assign the "Bonaerense" to the late Pleistocene (Fidalgo
et al, 1975); however, Frenguelli (1957) and Ameghino (1908)
suggested a middle-late Pleistocene and an early Pleistocene
age, respectively, for this unit. Pascual and Fidalgo (1972)
and Marshall and Pascual (1978) correlated the "Bonaerense"
with the late Pleistocene on basis of its fossil mammal fauna.

Fidalgo et al (1975) indicated that from the geological point
of view it is not possible to distinguish all the sedimentary
units from the Lower Pliocene to the Upper Pleistocene in
Buenos Aires Province; however, they have observed
unconformities between Tertiary and Quaternary sediments.

Samples for this study were collected from the walls of two
excavations in the City of Buenos Aires (A) and one in the
City of La Plata (B; Figure 1).

The two excavations in the City of Buenos Aires were situated
at different topographic levels. The relationship between these

11

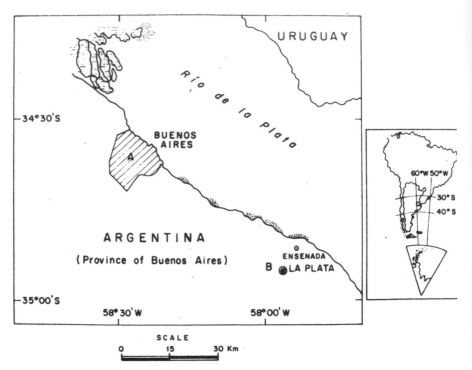

Figure 1. Locations of sampling sites; A: City of Buenos Aires
and B: City of La Plata, Buenos Aires Province. Samples are
collected from excavations carried out for building foundation.

and the depths of these excavations was such that it was
possible to sample a continuous sequence 23 m thick of
sediments assigned to the "Bonaerense" and "Ensenadense". In
this city it is not easy to define accurately the "Ensenadense"-
"Bonaerense" transition (González Bonorino, 1965). This
transition is defined in the sequence of sediments exposed in
the excavation situated at the higher stratigraphic level
(Valencio and Orgeira, 1983). However, the depth of this
transition is not precise; these authors suggested that this
transition is situated at about 10,5 m below ground level. In
the other excavation only sediments assigned to the
"Ensenadense" were found (Nabel and Valencio, 1981). The
lithology of the sediments of the "Ensenadense" and
"Bonaerense" of these two excavations is shown in Figure 5.
Samples of the "Ensenadense" and "Bonaerense" from La Plata
were collected from the walls of the excavation, about 30 m

12

deep, carried out for the foundation of the Nuevo Teatro Argentino (B; Figure 1). The "Ensenadense"-"Bonaerense" boundary
is there at 8 m below ground level (F. Fidalgo, Universidad
Nacional de La Plata, personal communication). The lithology
of these sediments (Figure 6) is remarkably different from
that of the sequence of Buenos Aires. On the basis of the
lithology the sequence of the "Ensenadense" and "Bonaerense"
from La Plata was subdivided into 15 geological sections
(A through N; Bobbio, 1983, and Devicenzi, 1983). The boundary
between the sections J and K, at about 14 m below ground level
and within the "Ensenadense", is an unconformity (F. Fidalgo,
personal communication).

SAMPLING

The "Ensenadense" and "Bonaerense" sediments were sampled
using plastic cylinders (0.025 m in diameter; 0.025 m in
height). They were introduced into the walls by hand or using
special tools according to the hardness of the sediment.

   In Buenos Aires, samples were collected from two excavations
(item 2). In the excavation situated at the higher topographic
level, 129 samples were obtained from a sedimentary sequence
14.5 m thick (Valencio and Orgeira, 1983). In the excavation
situated at the lower topographic level, 119 samples were
collected from a sequence 10 m thick of sediments of the
"Ensenadense" (Nabel and Valencio, 1981). Most of the samples
were recovered from different stratigraphic positions. The
mean stratigraphic interval between the axes of two consecutive
cylinders was 0.10 m; that is, the mean of the unsampled
stratigraphic interval between two consecutive cylinders was
0.075 m. Several samples were also taken at the same
stratigraphic level in some sections of the sequence in order
to test the consistence of the paleomagnetic data.

   In La Plata, 228 samples were collected from a sedimentary
sequence 28 m thick. Sampling was carried out following the
same technique used in Buenos Aires. The mean stratigraphic
distance between the axes of two consecutive cylinders was
0.12 m (Bobbio et al., 1985).

The intensity of natural remanent magnetization (NRM) was measured using a fluxgate slow speed spinner magnetometer (Vilas, 1979). Susceptibility measurements were made using a balanced double coil susceptibilimeter type RMSH III manufactured at the Tata Institute, Bombay. The stability of the NRM was investigated by alternating field (AF) demagnetization. Pilot samples were demagnetized successively in 25, 50, 75, 100, 125, 150, 175, 200, 225, 250 and 300 Oe; some of them were also demagnetized in 350 and 400 Oe. NRM directions of most of the samples showed either no systematic change on AF demagnetization or a viscous magnetization which could easily be removed by AF demagnetization in 25 or 50 Oe. Median destructive fields (MDFs) fell in the range 75-200 Oe; samples with higher MDF values were observed to exhibit smaller changes in NRM direction. Typical demagnetization curves for the "Ensenadense" and "Bonaerense" from Buenos Aires and La Plata are shown in Figure 2. Bulk demagnetization of not-pilot samples was carried out in 150-200 Oe peak fields.

The directions of the NRM and stable remanence for the collected samples from Buenos Aires and La Plata are shown in Figure 3 and 4, respectively. On the basis of the latter directions, virtual geomagnetic poles (VGPs) for each of these samples were calculated. Two mean poles were computed from these VGPs; one for samples from Buenos Aires and the other for samples from La Plata. VGPs situated more than 40° from these mean poles were rejected and two new mean poles were calculated. In this way, two populations of VGP's were obtained all within 40° of the mean poles for the samples from Buenos Aires and La Plata, respectively. These populations of VGP's yield two paleomagnetic poles for the "Ensenadense" and "Bonaerense": one for Buenos Aires at lat. 88.7° S, long. 254.7° E (N= 154, K= 10.3, A_{95}= 3.6°) and the other for La Plata at lat. 88° S, long. 264° E (N= 136, K= 12, A_{95}= 3.5°). The rejected VGP's were classified as oblique.

The magnetic stratigraphies for the "Ensenadense" and "Bonaerense" from Buenos Aires and La Plata, consisting of logs of stratigraphic level-value of cleaned declination (D) and inclination (I), are plotted alongside the logs of distances of VGP's from the mean polar positions in Figures 5 and 6, respectively. The distance of a VGP from the mean pole defines the polarity of the stable remanence of the sample: 0-40°,

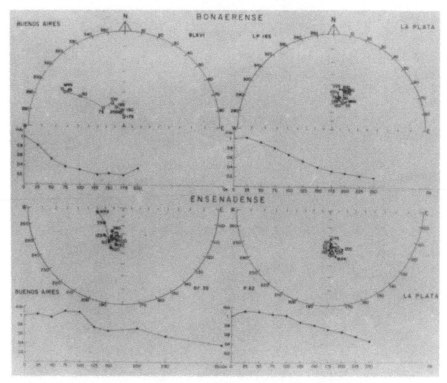

Figure 2. Changes in direction and intensity of magnetization
by progressive AF cleaning for one specimen of the "Bonaerense"
and the "Ensenadense" from each of the sampling sites. Solid
symbols indicate downward dipping directions.

normal polarity; 40°-140°, oblique polarity, and 140°-180°,
reversed polarity. The polarity of the stable remanence of
samples defines the magnetostratigraphy of each sequence of
sediments (Figures 5 and 6).

Logs of stratigraphic level-value of intensity of NRM (Jn)
and magnetic susceptibility (χ) are also plotted for the
"Ensenadense" and "Bonaerense" sedimentary sequences throughout
the studied excavations. The patterns of these logs for the
same excavation show a remarkable similarity, though they
differ from one excavation to another (Figures 5 and 6).

The magnetic susceptibility is not related to the intensity
of the geomagnetic field during the physicochemical processes
in which sediments acquire their NRM. On the other hand, it is

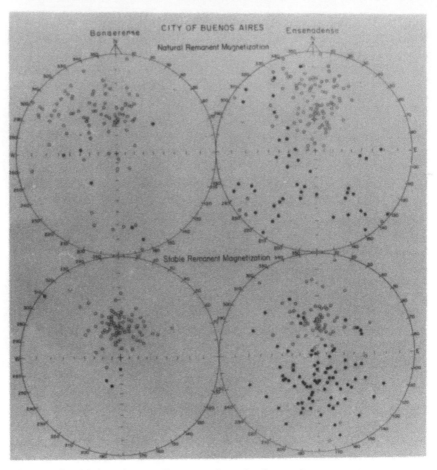

Figure 3. Directions of natural and cleaned remanent magnetization for "Ensenadense" and "Bonaerense" specimens from Buenos Aires. The direction of the present geomagnetic field is shown by ⊕. The key of the symbols is the same given in Figure 2.

related to the number, chemical composition and/or size of their magnetic minerals. Therefore, the comparison of logs of χ and Jn for each excavation indicates that their variations in time are due to variations in the number, chemical composition and/or size of the accumulated magnetic minerals.

THE REVERSAL TIME SCALE FOR THE LATE CENOZOIC

In the last years, new polarity events of the geomagnetic field

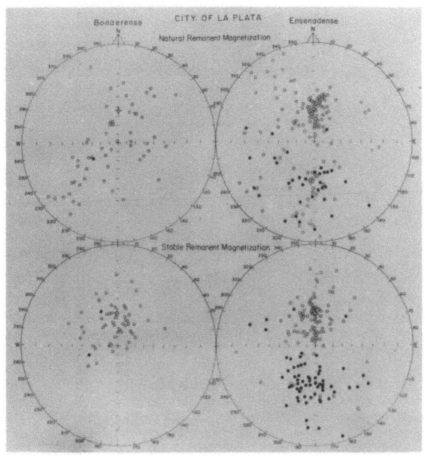

Figure 4. City of La Plata. Directions of natural and stable remanent magnetization for "Ensenadense" and "Bonaerense" specimens. The key of the symbols is the same given in Figure 2.

within the Brunhes Normal and Matuyama Reversed Epochs have been proposed.

These events have not been accepted unanimously by the paleomagneticians. The most important purpose of this paragraph is mainly to review and discuss data related with the existence of normal events in the lower Matuyama Reversed Epoch, between the base of the Olduvai Event and the Gauss-Matuyama Epochs transition (Figure 7). For the Gauss-Matuyama boundary, an age of 2.41 ± 0.01 Ma (McDougall and Aziz-Ur-Rahman, 1972) is accepted; this was derived from K/Ar ages calculated by means

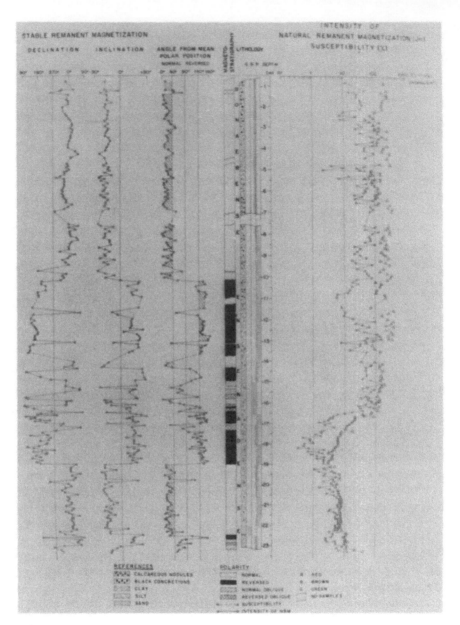

Figure 5. City of Buenos Aires. Declination and inclination
of stable remanence and angle between the VGP for each
stratigraphic level and the paleomagnetic pole through the
"Ensenadense" and "Bonaerense" sequence. The magnetostratigraphy, the lithology, and the intensity of natural
remanent magnetization (J_n) and susceptibility (χ) logs for
this sequence are also shown.

of ^{40}K decay constant proposed by Aldrich and Wetherill (1958; quoted in McDougall, 1979).

Data for the time scale of reversals of the geomagnetic field in the late Cenozoic are provided by: 1) oceanic magnetic anomalies; 2) the K/Ar age and the polarity of the stable magnetic remanence of subaerial volcanic rocks; 3) the magnetostratigraphy of sequences of rocks exposed on continents and dated by fossil assemblages and other techniques, and 4) the magnetostratigraphy of sequences from deep-sea cores. We must keep in mind that short polarity events of the geomagnetic field, on account of various causes, are not recorded by rocks of roughly the same age exposed in different sites.

Oceanic magnetic anomalies do not determine by themselves the reversal time scale because, frequently, the age of the magmatic rocks from the bottom of the oceans is unknown. But, when they are calibrated against known points of the radiometric time scale, the oceanic magnetic anomaly profiles provide a nearly continuous record of the geomagnetic field. Heirtzler et al. (1968) reported a magnetic anomaly (X anomaly), associated with a normal polarity remanence narrow zone, slightly older than the Olduvai Normal Event (Figure 7). This was probably the first evidence of a short normal within the lower Matuyama. Emilia and Heinrichs (1972) indicated an average age of 2.3 Ma (standard deviation 0.1 Ma) for the X anomaly. McDougall (1979) quoted an age of 2.17 Ma for this anomaly, whereas Rea and Blakeley (1975) suggested an age of 2.25 Ma.

The time scale for the late Cenozoic reversals of the geomagnetic field achieved on basis of the K/Ar age and the magnetic polarity of subaerial rocks requires precisely radiometric dates. However, it is not possible to obtain a complete time scale of reversals using this technique because of the episodic character of the volcanic activity. Valencio et al. (1970a, 1970b) reported stable remanence of normal polarity for two basaltic lava flows from the extra-Andean area of Neuquén Province (Argentina), which yielded K/Ar ages within the early Matuyama (2.30 ± 0.15 Ma and 2.31 ± 0.09 Ma). The standard deviations for these radiometric ages are larger than those quoted by other authors for rocks of equivalent age. However, the paleomagnetic data and the K/Ar age for the latter lava flow imply one normal event in the lower Matuyama Reversed

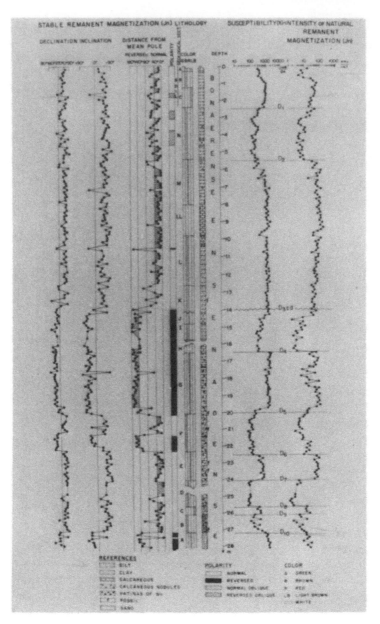

Figure 6. City of La Plata. Declination and inclination of
stable remanence and angle between the VGP for each
stratigraphic level and the paleomagnetic pole through
"Ensenadense" and "Bonaerense" sequence. The magnetostratigraphy, the lithology, and the intensity of n.r.m (Jn)
and susceptibility (χ) logs for this sequence are also
shown.

20

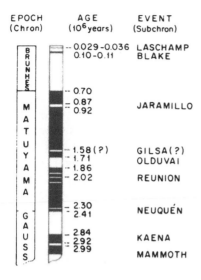

EPOCH (Chron)	AGE (10^6 years)	EVENT (Subchron)

Normal

Reversed

Figure 7. The reversal time
scale for the late Cenozoic
adopted in this paper.

Epoch (Valencio et al., 1975) if the standard deviation of the
date is accepted as an absolute error and an estimate of
2.41 Ma is accepted for the Gauss-Matuyama polarity transition.

The paleomagnetism of sequences of sediments exposed on
continents provides another source of information about short
polarity events. However, the record of short polarity events
can be obscured due to stratigraphic gaps, variations in the
rate of deposition and delays in the time between deposition
and magnetization of sediments. The paleomagnetic chronology
for a Pliocene-early Pleistocene sequence of marine sediments
from New Zealand revealed three events of normal polarity
within the lower Matuyama Epoch (Kenneth and Watkins, 1971).
Shuey et al. (1974) and Brown et al. (1978) reported four or
possibly five such zones in sediments from southwestern
Ethiopia which could be interpreted as short normal events in
early Matuyama time (one is nearly 2.32 Ma). Kochegura and
Zubakov (1978) reported two normal polarity events at 2.13 and
2.30 Ma, respectively, recorded in marine sediments of the

21

USSR (the oldest was dated by the fission-track method).
Liddlicoat **et al.** (1980) reported a zone of normal-magnetic
inclination near the beginning of the Matuyama Epoch from the
study of one 930 m long core from California, USA. Kristjansonn
et al. (1980) reported two normal events which occurred in the
lower Matuyama in Pliocene and Plio-Pleistocene sequences of
volcanic rocks from Iceland. Christoffel and Mak (1981)
reported normal polarity events occurred near the Gauss-
Matuyama transition in sequences of sediments from New Zealand
that were deposited on the continental shelf or shelf slope.

Briefly, information from oceanic magnetic anomalies,
sequences of sediments and subaerial lava flows strongly
suggests that, at least, a short event of normal polarity
occurred in early Matuyama times. These data are not conclusive
to indicate the precise age of that event. As the different
techniques agree in suggesting a normal polarity event at
2.3 Ma, this age is tentatively suggested for that event. The
late Cenozoic events of the geomagnetic field have been
usually named after the name of the collecting sites for the
paleomagnetic studies. Consequently, the name of Neuquén Normal
Event has been suggested for the normal event at the base of
the Matuyama Reversed Epoch (Valencio, 1981).

The time scale of the late Cenozoic reversals used in this
paper is shown in Figure 7.

INTERPRETATION OF RESULTS

The key to establish the correlation tie-lines of the magnetostratigraphy for a sequence of rocks and the time scale of
reversals of the geomagnetic field is to have, at least, one
reliable link between them. In this case, this link is the
"Bonaerense", the youngest unit of the "Pampeano" sediments,
because there is a general agreement in correlating it with
the Upper Pleistocene (Brunhes magnetic age; normal polarity
NRM). Therefore, we started the correlation of the magneto-
stratigraphies for the sequences from Buenos Aires and La Plata
and the time scale of reversals for the late Cenozoic from the
top of these sequences.

The correlation tie-lines adopted for the sequence of the
"Ensenadense" and "Bonaerense" sediments from Buenos Aires
are shown in Figure 8. They suggest a predominantly Brunhes
magnetic age (< 0.7 Ma; middle to late Pleistocene) for the

"Bonaerense". However, at the base of the "Bonaerense" (from 10.15 to 10.35 m; Figure 5) three samples are carriers, from the top to the base, of oblique normal and reversed polarity (two samples) remanences. This suggests a late Matuyama magnetic age (> 0.7 Ma) for the latter two samples. However, we should remind here that the transition "Ensenadense"- "Bonaerense" is not precisely defined in the studied excavation; that is, these three samples might have been erroneously assigned to the "Bonaerense".

The remanence of sediments of the section of the "Ensenadense" from 10.40 m to 19.00 m is of predominantly reversed polarity; therefore, they were correlated with the Matuyama Reversed Epoch. The normal polarity subsections within this predominantly reversed section were correlated with the subchrons within the Matuyama chron (Figure 8).

Finally, the sediments of the "Ensenadense" section from 19.00 m to 23.00 m have remanence of predominantly normal polarity; at the base of this section, some samples are carriers of remanence of oblique polarity and one of them, reversed polarity. The sediments of this section were correlated with the late Gauss Chron and the Kaena Subchron, respectively. That is, our interpretation suggests a late Gauss to Matuyama (2.84-> 0.7 Ma, late Pliocene to middle-late Pleistocene) magnetic age for the sequence of "Ensenadense" sediments from Buenos Aires.

Briefly, the magnetic age for the "Bonaerense" from this site is consistent with the age suggested for this unit by Frenguelli (1975) and older and partially consistent with that assigned to the Lujanense Mammal Age (Table 1). The magnetic age for the "Ensenadense" is partially consistent with the ages suggested for this unit by Ameghino (1908) (Pliocene) and Frenguelli (1975) (middle Pleistocene) and older and partially consistent with that assigned to the Ensenadense Mammal Age (Table 1).

The correlation tie-lines adopted for the sequence of the "Ensenadense" and "Bonaerense" from Buenos Aires and the time scale of reversals (Figure 8), suggest two rates of accumulation: 14 mm/1000 years for the upper 13 m and 7.5 mm/ 1000 years for the lower 10 m of sediments. They also suggest a discontinuity in the accumulation of sediments at about 13.15 m. Sharp changes in the lithology, the intensity of NRM and magnetic susceptibility and a change of polarity of the geomagnetic field occur at this depth (Figure 5). These are

23

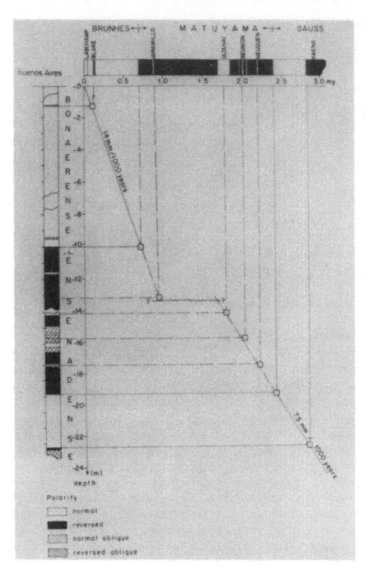

Figure 8. City of Buenos Aires. Correlation tie-lines adopted
for "Ensenadense" and "Bonaerense" sequence and the late
Cenozoic time scale of reversals.

interpreted as evidence of a discontinuity in the accumulation
of sediments, as it will be discussed later. That is, two
different lines of reasoning agree in suggesting a
discontinuity in the Buenos Aires "Ensenadense" sediments.

24

The magnetostratigraphy for the "Ensenadense" and "Bonaerense" sediments from La Plata is rather different from that of Buenos Aires (Figures 6 and 5, respectively). There are also remarkable differences in the lithology and the patterns of the intensity of NRM and magnetic susceptibility logs. As it has been mentioned (item 2), the La Plata sequence was subdivided into 15 geological sections on basis of their lithology. Particularly, field geological data indicate that the boundary between the lithologic sections J and K is an unconformity (d). This unconformity is also coincident with changes in the values of the intensity of NRM and magnetic susceptibility, and a reversal of the geomagnetic field, from reversed to normal polarity (D_3). It is difficult to explain that sharp changes in the lithology and in the number, chemical composition and/or size of magnetic mineral of sediments are coeval with a reversal of the geomagnetic field because their causes are entirely different. We interpret this as an evidence of a discontinuity in the sediment accumulation; that is, the lithology and the scalar and vectorial magnetic variables agree with the field geology defining a discontinuity in the sediment accumulation at the J-K transition ($d=D_3$).

Three other coincident sharp variations in the lithology and in the scalar and vectorial magnetic variables are defined in the sequence of La Plata "Ensenadense" sediments. They have been named D_5, D_6 and D_{10} (about 20.0; 22.5 and 27.0 m deep, respectively); the latter is coincident with an irregular surface between the lithological sections A and B. They are also interpreted as discontinuities in the sediment accumulation; perhaps, some of them can be minor unobservable unconformities. Coincident variations in the lithology and the scalar magnetic variables are also defined in the sequence; they have been named D_1, D_2, D_4, D_7, D_8 and D_9. They could be evidence of minor discontinuities in the accumulation of the "Ensenadense" and "Bonaerense" sediments in La Plata as well.

Briefly, field geological data, the lithology and the scalar and vectorial magnetic variables suggest that the accumulation of the "Pampeano" sediments from La Plata may have been interrupted several times during periods of unknown duration. This constitutes a serious difficulty for the correlation of the magnetostratigraphy of these sediments and the time scale of reversals.

The two patterns of correlation tie-lines adopted for the sequence of the La Plata "Ensenadense" and "Bonaerense"

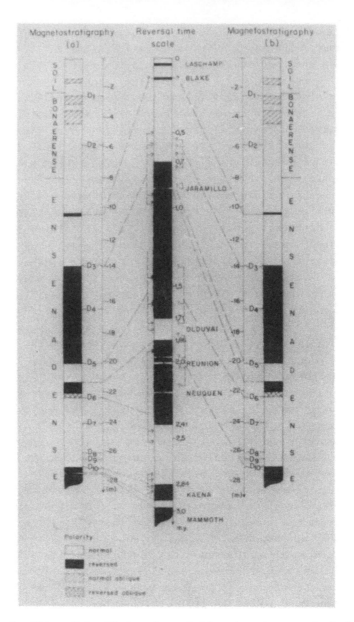

Figure 9. City of La Plata. Correlation tie-lines adopted for the sequence of the "Ensenadense" and "Bonaerense" and the late Cenozoic time scale of reversals. Two interpretations are suggested.

sediments are shown in Figure 9. They roughly suggest a Brunhes magnetic age ($<$ 0.7 Ma) for the "Ensenadense" and "Bonaerense" sediments from the higher 14 m of the sequence and a Matuyama magnetic age ($>$ 0.7 Ma) for the section of the "Ensenadense" sediments, of predominantly reversed remanence, located between 14 and 22.5 m. The hiatus associated with the unconformity **d** and the magnetic discontinuity D_3 indicates the absence of an unknown thickness of sediments of early Brunhes and late Matuyama magnetic age. This allows us to define more precisely the age of the "Ensenadense" and "Bonaerense" sediments of predominantly normal polarity remanence from the top of the sequence. In fact, it suggests a middle-late Bruhnes age for these sediments. Therefore, the reversed polarity event recorded at 10.5 m could be correlated with the Blake Subchron (Figure 9).

The correlation of the "Ensenadense" sections of predominant reversed (14.0 - 22.5 m) and normal polarity (22.5 - 28.5 m) remanence and the time scale of reversals depend on the time span of the hiatus associated with the unconformity d= D_3 and the discontinuities D_5, D_6 and D_{10}. The two patterns of correlation tie-lines adopted are shown in Figure 9.

The interpretation (a) suggests: 1) magnetic Matuyama age ($>$ 0.7 Ma; $<$ 2.41 Ma) for the "Ensenadense" sediments from 14.0 to 22.5 m deep, and 2) magnetic late Gauss age (2.41 Ma; $<$ 3.0 Ma) for the "Ensenadense" sediments from 22.5 to 28.5 m. For the entire sedimentary sequence, this interpretation suggests middle-late Pleistocene age for the "Bonaerense" and . late Pliocene to middle-late Pleistocene age for the "Ensenaden-se". The magnetic age for the "Bonaerense" is consistent with the age suggested for this unit by Frenguelli (1957), and older and partially consistent with that assigned to the Lujanense Mammal Age (Table 1). The magnetic age for the "Ensenadense" is partially consistent with the ages suggested for this unit by Ameghino (1908; Pliocene) and Frenguelli (1957; middle Pleistocene), and older and partially consistent with those assigned to the Ensenadense Mammal Age (Table 1).

The interpretation (b) suggests magnetic Matuyama age ($>$ 0.7 Ma; $<$ 2.41 Ma) for the "Ensenadense" sediments from 14.0 to 28.5 m deep. For the whole sedimentary sequence, this interpretation suggests a middle-late Pleistocene age for the "Bonaerense" and a late Pliocene to middle-late Pleistocene age for the "Ensenadense". The magnetic age suggested for the "Bonaerense" in this interpretation is similar to that suggested

in the former interpretation (a). The magnetic age suggested
for the "Ensenadense" is partially consistent with the ages
suggested for this unit by Ameghino (1908) and Frenguelli (1957)
(Table 1) and older and partially consistent with those assigned
to the Ensenadense Mammal Age.

DISCUSSION

The geology, lithology and the paleomagnetic and magnetic
susceptibility data for the "Ensenadense" and "Bonaerense"
sediments from the Cities of Buenos Aires and La Plata indicate
that they were accumulated under different geological processes.
    In Buenos Aires, the accumulation of these sediments was
practically continuous; only one period of no sedimentation
and/or erosion, recorded at about 13.15 m below ground level,
interrupted the accumulation of the "Ensenadense" sediments.
The magnetostratigraphy for this sedimentary sequence suggests
a predominantly Brunhes (> 0.7 Ma) magnetic age for the
"Bonaerense" and a Matuyama to late Gauss magnetic age
(> 0.7 Ma - 2.84 Ma) for the "Ensenadense".
    In La Plata, the accumulation of the "Ensenadense" and
"Bonaerense" sediments was discontinuous. This is suggested
by one unconformity and four discontinuities in the lithology
and the scalar and vectorial magnetic variables; particularly,
one of these discontinuities is coincident with the
unconformity. Six discontinuities in the lithology and scalar
magnetic parameters could also be associated with minor periods
of no accumulation. The magnetostratigraphy for the La Plata
"Ensenadense" and "Bonaerense" sediments admits two
interpretations, which suggest a magnetic age middle-late
Brunhes for the "Ensenadense" and "Bonaerense" sediments from
the highest 14.0 m of the sequence. One of the interpretations
suggests magnetic late Gauss-Matuyama age (< 3.0 Ma; > 0.7 Ma)
for the "Ensenadense" sediments from the lowest 14.0 m of the
sequence. The other interpretation suggests magnetic age
Matuyama (< 2.41 Ma; > 0.7 Ma) for these "Ensenadense"
sediments.
    The magnetic age for the "Ensenadense" from the Cities of
Buenos Aires and La Plata embraces, roughly, the time-span
late Pliocene middle-late Pleistocene. This is partially
consistent with the ages suggested for this unit by Ameghino
(1908; Pliocene) and Frenguelli (1975; middle Pleistocene;

Table 1); however, it suggests a longer period of accumulation
for the "Ensenadense" sediments. Particularly, the magnetic
late Pliocene age is older than the Land Mammal Age suggested
for this unit (Table 1). The significance of this
interpretation concerning the age of the establishment of a
stable land-bridge between South and North America is
discussed elsewhere (Valencio, 1985).

ACKNOWLEDGEMENTS

The authors wish to thank the Universidad de Buenos Aires, the
Comisión de Investigaciones Científicas de la Provincia de
Buenos Aires (CIC) and the Consejo Nacional de Investigaciones
Científicas y Técnicas (CONICET) for the support. They also
thank the useful suggestions from Dr F. Fidalgo and
Dr R. Pascual (Universidad Nacional de La Plata) and
Dr O. Schneider for his assistance in editing this paper.

REFERENCES

Ameghino, F. 1889. Contribución al conocimiento de los mamí-
    feros fósiles de la República Argentina. **Acad. Nac. Ciencias,**
    6:1-1027, Córdoba.
Ameghino, F. 1908. Las formaciones sedimentarias de la región
    litoral de Mar del Plata y Chapadmalal. **Museo Nacional de**
    Historia Natural, Anales, Serie 3, 10:343-428. Buenos Aires,
    Argentina.
Bobbio, M.L. 1983. Estudio geológico y paleomagnético de los
    sedimentos del Ensenadense aflorantes en la excavación
    realizada para la construcción del Nuevo Teatro Argentino,
    Ciudad de La Plata. Tesis de Licenciatura, Departamento de
    Ciencias Geológicas, Universidad de Buenos Aires, unpublished.
Bobbio, M.L., Devincenzi, S.M., Orgeira, M.J. and Valencio, D.A.
    1985. La magnetoestratigrafía de una secuencia de sedimentos
    del Ensenadense y Bonaerense de la Ciudad de La Plata (Exca-
    vación Nuevo Teatro Argentino): su significado geológico.
    Rev. Asoc. Geol. Arg. in press.
Brown, F.H., Shuey, R.T. and Croes, M.K. 1978. Magneto-
    stratigraphy of the Shungura and Usno Formations, south-
    western Ethiopia: new data and comprehensive reanalysis.
    Geophys. J.R. astr. Soc., 54:519-538.
Butler, R.F., Marshall, L.G., Drake, R.E. and Curtis, G.H.

1984. Magnetic polarity stratigraphy and ^{40}K-^{40}Ar dating of Late Miocene and Early Pliocene Continental Deposits, Catamarca Province, NW Argentina. **Journal of Geology.** 92:623-636.

Christoffel, D.A. and Mak, W. 1981. Magnetic reversal stratigraphy from paleomagnetic measurements on sedimentary sequences in New Zealand. **IAGA Bulletin** N° 45, 215.

Devincenzi, S.M. 1983. Estudio geológico y paleomagnético de los sedimentos del Ensenadense-Bonaerense aflorantes en la excavación realizada para la construcción del Nuevo Teatro Argentino, Ciudad de La Plata. Tesis de Licenciatura. Departamento de Ciencias Geológicas. Universidad de Buenos Aires, unpublished.

Emilia, D.A. and Heinrichs, D.F. 1972. Paleomagnetic events in the Brunhes and Matuyama Epoch identified from magnetic profiles reduced to the pole. **Mar. Geophys. Res.,** 1:436-444.

Fidalgo, F, De Francesco, F. and Pascual, R. 1975. Geología superficial de la llanura bonaerense. **Relatorio V Congr. Geol. Argentino,** 103-138.

Frenguelli, J. 1957. Neozoico. **Geografía de la República Argentina. GAEA.** 3:1-218.

Gonzalez Bonorino, F. 1965. Mineralogía de las fracciones arcillas y limo del Pampeano en el área de la Ciudad de Buenos Aires y, su significado estratigráfico y sedimentológico. **Rev. Asoc. Geol. Argent.,** 20(1):67-150.

Hallam, A. 1972. Continental drift and the fossil record. **Scientific American,** November, p.57-66.

Heirtzler, R., Dcikson, G.O., Herron, D.M., Pitman III, N.C. and Le Pichon, X. 1968. Magnetic anomalies, geomagnetic reversals and motions of the ocean floor and continents. **J. Geophys. Res.** 73(6):2119-2136.

Kenneth, J.P. and Watkins, D.N. 1971. Paleomagnetic chronology of Pliocene-Early Pleistocene climates and the Pio-Pleistocene boundary in New Zealand. **Science,** 171:276-279.

Kochegura, V.V. and Zubakov, V.A. 1978. Paleomagnetic time scale of the Ponto-Caspian Plio-Pleistocene Deposits. **Palaeogeography, Palaeoclimatology, Palaeoecology.** 23:151-160.

Kraglievich, J.L. 1952. El perfil geológico de Chapadmalal y Miramar, Pcia de Buenos Aires. **Rev. Mus. Cienc. Nat. y Trad.** 1(1):8-37. Mar del Plata.

Kristjansonn, L., Fridleifsson, I.B. and Watkins, N.D. 1980. Stratigraphy and palaeomagnetism of the Esja, Eyrafjall and Akrafjall Mountains, S.W. Jaland. **Journal of Geophysies.** 47:31-32.

Liddicoat, J.C., Opdyke, N.D. and Smith, G.I. 1980. Palaeomagnetic polarity in 930 m core from Searleys Valley, California. **Nature**. 286:22-25.

McFadden, B.J., Siles, O., Zeitler, P. Johnson, N.M. and Campbell, Jr., K.E. 1983. Magnetic Polarity Stratigraphy of the Middle Pleistocene (Ensenadan) Tarija Formation of Southern Bolivia. **Quaternary Research**. 19:172-187.

Marshall, L.G. and Pascual, R. 1978. Una escala temporal radimétrica preliminar de las edades mamífero del Cenozoico medio y tardío sudamericano. **Obra del Centenario del Museo de La Plata**. 5:11-28.

Marshall, L.G., Butler, R.F., Drake, R.E., Curtis, G.H. and Tedford, R.H. 1979. Calibration of the Great American Interchange. **Science**. 204:272-279.

Marshall, L.G., Butler, R.F., Drake, R.E. and Curtis, G.H. 1982. Geochronology of Type Uquian (Late Cenozoic) Land Mammal Age, Argentina. **Science**. 216:986-989.

McDougall, I. 1979. The present status of the geomagnetic polarity time scale. **The Earth: its origin, structure and evolution**. McElhinny, M.W. (ed.), Academic Press, London (Publication N°1288, Research School of Earth Sciences, A.N.U., p.1-34).

McDougall, I. and Aziz-Ur-Rahman. 1972. Age of the Gauss-Matuyama boundary and the Kaena and Mammoth Events. **Earth Planet. Sci. Lett.** 14:367-380.

Nabel, P.E. and Valencio, D.A. 1981. La magnetoestratigrafía del Ensenadense de la Ciudad de Buenos Aires: su significado geológico. **Rev. Asoc. Geol. Argent.** 36(1):7-18.

Pascual, R., Ortega Hinojosa, E.J., Gondar, D. and Tonni, E. 1965. Las edades del Cenozoico mamalífero de la Argentina, con especial atención a aquellas del territorio Bonaerense. **An. Com. Invest. Cient. Prov. Buenos Aires.** 1(6):165-193.

Pascual, R. and Fidalgo, F. 1972. The problems of the Plio-Pleistocene boundary in Argentina (South America). **International Colloquium on the problem "The boundary between Neogene and Quaternary"**. Collection of papers, II, Moscow.

Rea, D.K. and Blakeley, R.J. 1975. Short wavelength magnetic anomalies in a region of rapid sea-floor spreading. **Nature**. 255:126-128.

Shuey, R.T., Brown, F.H. and Croes, M.K. 1974. Magnetostratigraphy of the Shungura Formation, Southwestern Ethiopia; fine-structure of the Lower Matuyama Polarity

Epoch. **Earth Planet. Sci. Lett.** 23:249-260.

Tarling, D. 1981. The geological evolution of South America during the last 200 million years. In Ciochon, R.C. and Chiarelli, A.B. (Eds). **Evolutionary Biology of the New World Monkeys and Continental Drift.** Plenum Press, p.1-41. New York.

Valencio, D.A. 1981. Evidence for a short normal event at the base of the Matuyama Reversed Epoch: Neuquén Normal Event. **IAGA Bulletin N°45,** p.215.

Valencio, D.A. 1985. New evidence about the age of the land between South and North America. **Journal of Geodynamics,** in press.

Valencio, D.A., Linares, E. and Creer, K.M. 1970a. Palaeo-magnetism and K-Ar ages of Cenozoic basalts from Argentina. **Geophys. J.R. astr. Soc.** 19:147-164.

Valencio, D.A., Linares, E. and Vilas, J.F. 1970b. On the age of the Matuyama-Gauss Transition. **Earth Planet. Sci. Lett.** 8(2):179-182.

Valencio, D.A., Vilas, J.F. and Mendia, J.E. 1975. Palaeo-magnetism of Quaternary rocks from South America. **An. Acad. Brasi. Cienc.** 47 (Suplemento):21-32.

Valencio, D.A. and Orgeira, M.J. 1983. La magnetoestratigrafía del Ensenadense y Bonaerense de la Ciudad de Buenos Aires: Parte II. **Rev. Asoc. Geol. Argent.** 38(1):24-33.

Vilas, J.F. 1979. El magnetómetro UBA, 8 Hz y su aplicación a los estudios paleomagnéticos. Tesis doctoral, Departamento de Física, Universidad de Buenos Aires, unpublished.

Webb, S.D. 1976. Mammalian faunal dynamics of the Great American Interchange. **Paleobiology.** 2:216-234.

MÓNICA C.SALEMME
Vertebrate Palaeontology Division, Museo de La Plata, Argentina, and CIC
LAURA L.MIOTTI
Archaeology Division, Museo de La Plata, Argentina, and CONICET

3

Zooarchaeology and palaeoenvironments: Some examples from the Patagonian and Pampean regions (Argentina)

ABSTRACT

The analysis of faunal remains from archaeological sites is interesting from different points of view. First, the exploitation of the fauna by the aborigines as an economic resource. Second, the interpretation of palaeoenvironments based on the modifications in the corology of different species since the late Pleistocene.

The aim of this paper is to analyze the palaeoenvironmental aspects and the exploitation of the fauna by hunters and hunter-gatherers using three archaeological sites from the Pampean and Patagonian regions as examples.

RESUMEN

El estudio de restos faunísticos procedentes de sitios arqueológicos es importante desde varios puntos de vista: 1) la explotación faunística por los indígenas como un recurso económico; 2) la interpretación de los paleoambientes basada en las modificaciones de la corología de diferentes especies desde el Pleistoceno tardío.

Este trabajo intenta analizar los aspectos paleoambientales y la explotación de la fauna por los cazadores y cazadores-recolectores, usando como ejemplos tres sitios arqueológicos de distintos ambientes, de las regiones Pampeana y Patagónica.

INTRODUCTION

The emphasis on the analysis of faunal remains from archaeological sites since 1970 is focused on the wide information provided by those remains about cultural-ecological adaptations of indigenous people (Payne, 1972; Clason, 1975).

This kind of analysis is interested not only in the archaeological knowledge but also in the biological research. In that sense, the faunal remains are considered from the archaeological point of view as an economic resource: a) as raw material (leather, bone, skin, feathers, tendons, etc.) in the aboriginal technology; b) as food resource (meat, marrow, fat, entrails, eggs) and c) as a luxury item or ritual element. These aspects comprise the main interest of Zooarchaeology (Olsen, 1971; Clason, 1975).

From the biological point of view, bone remains give information upon the modifications in the corology of different species through geological times and they may enable the scientist to reconstruct the palaeoecosystems or at least part of them. This is the aim of the Biologic Archaeology (Olson, 1982; Tonni, 1984).

Therefore, both disciplines do not differ in the studied object, although the approach to the problem is not exactly the same.

Taking into account the faunal remains from three archaeological sites, the goals of this contribution will concentrate on: 1) the palaeoenvironmental aspects and 2) the faunal exploitation by hunters.

The sites cited in this paper have been partially studied by the authors. The fauna recovered there provided relevant information to be analyzed from a zooarchaeological and palaeoenvironmental point of view. The faunal assemblages reviewed come from Los Toldos Cave 3 (Cardich **et al.**, 1973; Cardich, 1977; 1984; Cardich and Miotti, 1983); La Toma site (Salemme **et al.**, in press) and Cañada de Rocha site (Ameghino, 1880; Salemme, 1983). These examples are referred to the fauna that coexisted with the aborigines who lived during the Late Pleistocene-Holocene in the Pampean and Patagonian regions. Two of the sites were inhabited under different environmental conditions from the present ones and the third suggested some climatic oscillations which were detected through the faunal remains.

METHODOLOGICAL BACKGROUND

Some theoretical-methodological concepts will be developed in the following paragraph. They are illustrated in figures 1 and 2 and will be discussed and underlined in the text.

Considering the archaeological context in Butzer's (1980)

sense:

"... the goal of contextual archaeology should be the study of archaeological sites as part of human ecosystem, within which past communities interacted spacially, economically and socially with the environmental subsystem into which they were adaptatively networked" (Butzer, 1980; 417). So "This contextual approach heavily dependent on archaeobotany, zooarchaeology and geoarchaeology is new not in terms of its components but by virtue of its integrated, general goal of the human ecosystem" (1980:422).

Assuming a human occupation area (campsite, butchering site, workshop, etc) as part of the ecosystem, there are some taphonomic processes that act on a dead biologic community (**Thanatocoenosis**) and on the cultural remains (art and technofacts).

Taphonomy (**stricto sensu**) has been defined by Efremov (1940) as the "science of the laws of embedding" (**in** Morlan, 1980:30). Recently, the usage of this term included "all considerations bearing upon the passage of organic material from the biosphere to the litosphere..." (Morlan, 1980:30). So, this reduced assemblage of biological and non-biological pieces is called **Taphocoenosis** (Meléndez, 1977; Morlan, 1980; Wood and Johnson, 1978) (see figure 1).

The taphonomic processes have reduced the primitive ecological assemblage to a sample. This is well-documented in Klein & Cruz Uribe (1984) who postulated five basic stages that a sample suffers before being studied by scientist. Those processes could be, among others, the action of microorganisms, rodents, scavengers, chemical and physical factors, ants, earthworms, roots and anthropogenic activity in the past and at present times.

An archaeological sample gives us information about part of the way of life of extinct people which is obtained from an archaeological excavation (see Figure 2). The most important point of this paper is the faunal sample which refers only to the vertebrates (Figure 2).

The faunal sample represents a part of the living fauna in the area when it was inhabited by people. It is integrated by faunal remains incorporated into the sediments in two ways:
a) anthropogenic activity and b) natural deposition.

The anthropogenic activity is evidenced by:
1) overdimension of some species in the faunal sample;
2) intentional fracture and a great amount of splintered bones;
3) burned bones;

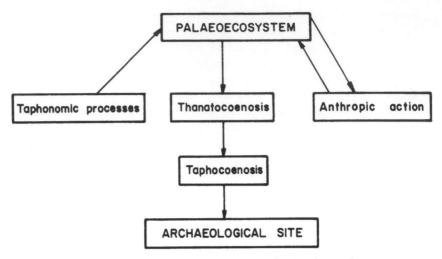

Figure 1. Taphonomic processes in the formation of an archaeological site.

4) bone artifacts;
5) selection of skeletal specimens;
6) bones, teeth, scutes, etc., as burial ornaments; and
7) faunal remains as part of the human burial assemblage.

   Although it is actually difficult to recognize whether some remains were deposited before or after the cultural event, bone remains from archaeological sites indicate different kinds of cultural activities.

   The overdimension of some species upon others, points out an anthropogenic selection. The reasons for that selection could be, among others, food preferences, taboos, domestication, resource availability, etc.

   This situation is specially verified in hunters and hunter-gatherers. In these groups, the cultural preferences are important as well as other factors, like the strategies employed to obtain the resources (Bettinger, 1978).

   The hunting activities are adapted to the preys; if they are social or lonely animals. According to these characteristics, the hunt will be single or collective. Different technics -for example, traps, nets, different projectile types, poisoning- would be used.

   Such assessment could be translated into an equation in which the relation is:

kind of prey $\longrightarrow$ hunting technics = different adaptative patterns
     (1)   (2)   (3)

where (2) is function of (1) and both of them condition (3).

Examples of hunters and resource-obtaining strategies were considered by different scientists not only for archaeological groups but also for ethnographical tribes. Among those mentioned, the extinct megamammals and the extant fauna were studied first. Although their ideas will not be discussed here, some authors should be cited: Flannery, 1967; Smith, 1975; Shimada, 1976; Jochim, 1976; Borrero, 1978; Binford, 1981; Correal Urrego, 1981; Foley, 1983; Mengoni Goñalons, 1983; Speth, 1983; Politis, 1984; among others.

In the case of herder and/or agriculturalist's societies, the overdimension of one species in the sample is specially evidenced by those related to the domestication processes (Raffino **et al.**, 1977; Cigliano **et al.**, 1973; Wing, 1972; 1977; Browman, 1974; Pollard, 1976; Davis, 1982; Grant, 1982).

The usefulness of animal resources selected by a human group could be inferred from the skeletal parts which constitute the sample; so, one can know if each prey was utilized partially or completely (meat, bone, fat, leather, feathers, eggs).

Examples of archaeological groups of Patagonia show that there was a tendency to use integrally the faunal resource, specially **Lama guanicoe** (Cardich and Miotti, 1983; Silveira, 1979). In the Pampean region, the animal resources were also wholly used, but two species were generally hunted: **Lama guanicoe** and **Ozotoceros bezoarticus** (Salemme and Tonni, 1983; Politis and Tonni, 1983; Fidalgo **et al.**, 1986). Besides, in both regions the ostrich ("ñandú": **Rhea americana**) has been a complementary but predilect resource (Falkner, 1974; Crawford, 1976; Priegue, 1971).

The "guanaco" (**L. guanicoe**) has been the main resource among the Patagonian hunters since the Late Pleistocene. Thus, the archaeological data indicate an entire usefulness of the faunal resource by groups who lived before the historic Tehuelches (i.e., the XVIIth century).

The archaeological data which demonstrate the usefulness of the animal resources are bone remains (different skeletal pieces), burned bones, shell-egg fragments, instruments such as scrapers, projectile points, knives and sometimes, pieces of leather. All of this information may be reinforced with the ethnohistorical data, that is the chronicles of the first travellers (i.e. Darwin, 1846; Pigafetta, 1894; Musters, 1911; Moreno, 1969; Gusinde, 1931; Outes, 1928; Priegue, 1971; Oviedo & Valdez, 1851).

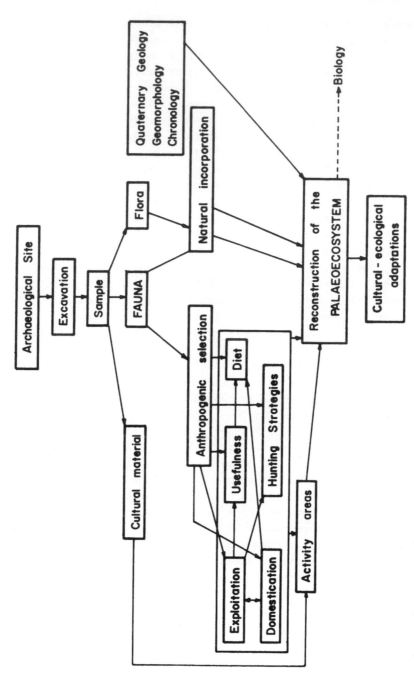

Figure 2. Processing and Interpretation of Data.

Before the introduction of the horse (**Equus caballus**), the Tehuelches used the "guanaco" for food, tents, clothes and medical purposes. Some of these necessities were replaced by the use of horses just after the beginning of the XVIIIth century.

The indicators of the diet in archaeological sites that have been inhabited by hunters are not only faunal remains but also material elements which indicate the use of the animal resources. The biggest energetic contribution in these societies comes from the animal proteins and fat contents (Service, 1966; Binford, 1981; Speth, 1983). The main evidences showing that the meat, fat, eggs, marrow and entrails were eaten are, for instance, cutmarks upon the bones, broken bones to obtain the marrow, pieces of shell-eggs and the presence of selected skeletal parts, at least those which provide a certain volume of meat.

Ethnohistorical information of Patagonian groups yielded numerous data about the use of meat, marrow and fat as main elements in the diet (Priegue, 1971). On the other hand, this kind of information is obtained from an archaeological sample by analizing the skeletal parts recorded, as it was already mentioned.

In the case of hunter-gatherers, the nutritional value is given proportionally by animal and vegetal products. Some voyagers noticed the gathering of edible fruits and roots (Musters, 1911). However, the direct archaeological evidence is not enough to confirm the use of vegetables in the diet.

The archaeological indicators in hunter-gatherer societies will be animal and vegetable remains (direct evidence) and/or technofacts (indirect evidence) (Aschero, 1985) which reflect the incorporation of both resources in a relatively equilibrated way.

The diet of archaeological groups could also be inferred through the analysis of human coprolites (Bottema, 1975), i.e. dehydrated dregs.

Some inferences about the activities developed in the site -activities areas- could be done if the entire record and information recovered in an archaeological excavation (sample) are considered. Following Sivertsen (1980), the sites could be classified as kill sites, butchering sites, campsites as it is indicated by the remains of the hunter-gatherers left within them.

In this case, it must be taken into account, not only the faunal remains, but also the cultural elements obtained and the areal geomorphology.

The faunal sample becomes essential to deduce some activities performed in the site. For example, the presence of certain skeletal parts will indicate that the preys were killed, or processed and/or eaten at the site (Borrero et al., 1985).

In that sense, in a processing and consuming site, the bone remains that may be found are the zeugopodials, carcasses or parts of them and probably, some cranial pieces like the mandibles. Likewise, a high percentage of broken bones can be found; they are produced by the intentional break to extract the marrow and/or to make bone instruments (Cardich and Miotti, 1983; Cardich and Laguens, 1984; Salemme et al., in press).

In a kill site, the skeletal parts that could be found are: autopodials (acropodials, metapodials, basipodials), vertebrae and cranial skeleton, for they are the skeletal parts with less volume of meat (Palanca and Politis, 1979; Speth, 1983; Politis, 1985). The useful cranial parts (muscle, glande and organs) and the encephalic mass could have been consumed in the same place where the preys were killed. This custom was observed in ethnographic groups (Gusinde, 1931; Pigafetta, 1894).

Both kinds of sites may be observed in the same settlement area, where an evidence of activities developed in a campsite, such as technofacts or food cooking (hearths), is registered. Los Toldos Cave 3 (see Figure 3) is an example of these characteristics (Cardich et al., 1973). The faunal sample showed that the individuals could have been hunted in the dell and/or the high plateaus, near the cave and carried into it in one single piece. Butchering and consumption labors were performed within the cave.

CLIMATE AND BIOGEOGRAPHY

A brief characterization of the environments in the chosen areas is necessary to understand the present ecosystems. These data plus the archaeological information recorded in the way explained in the precedent paragraph, let us interpret the palaeoenvironmental conditions.

The environments where the selected sites are located in, correspond to 1) the central plateau of Santa Cruz province; 2) the neighbourings of Sierra de la Ventana and, 3) the Pampa Ondulada, both in Buenos Aires province (see Figure 3).

Los Toldos Archaeological Locality is located south of the middle Deseado River (Santa Cruz province). This locality is composed of fifteen caves along a dell (1 km approximately). The data used in this paper come from the Cave 3, which is

situated at lat. 47° 22' S and long. 68° 58' W and 300 m
a.s.l. (see Figure 3).

Physiographically, the region corresponds to the extraandean
Patagonia; it is located in the Santa Cruz central plateau,
which is interrupted only by dells and basins with temporary
creeks. They constitute microenvironments which offer a better
refuge from the western and southern winds; thus, here, the
humidity concentration is higher and variability and frequency
of vegetation are richer than on the plateaus.

The cold, arid and windy climate corresponds to the arid-
semiarid conditions of extraandean Patagonia. The mean annual
temperature is 9.7° C; the mean annual rainfall is 161 mm (see
Figures 4 and 5). The predominant winds come from the west all
the year round. In winter, frosts are frequent, beginning in
April and going on until September. According to Thornthwaite's
classification, the climate is E_1 B_1' d b_4': arid, mesothermal,
with low or null water deficiency and summer concentration of
the thermal efficiency = 49% (O. Martinez, pers. com.).

Phytogeographically, this region corresponds to the
Patagonian steppe (Parodi **et al.**, 1947), with very xeromorphic,
low and open vegetation. There are well-watered and very
fertile, flat lowlands, with ciperaceous and gramineous
vegetation.

Zoogeographically, the region is within the Patagonian
Dominion (Cabrera and Yepes, 1960). The "guanaco" is the more
representative megamammal of the autochtonous fauna; the "puma"
(**Felis concolor**), several fox species (**Canis** (**Dusicyon**) **sp**)
and smaller felids (**Lynchailurus colocolus, Oncifelis geoffroyi**
and **Notoctifelis guinga**) are frequent as well.

The rodents are common in this area, predominantly the
families Caviidae and Cricetidae.

The "ñandú" or "choique" (**Pterocnemia pennata**) is a
flightless characteristic bird. Other birds which represent the
Patagonian avifauna are the "martinetas" (**Eudromia elegans**),
with several subspecies.

The Pampean region has been subdivided into several areas
(Daus, 1946) according to the physiographical characteristics.
The sites selected for this contribution are located in two of
those areas, presenting very sharp differences between them.

La Toma site is localized at lat. 38° 17' S and long. 61°
41' 40" W and 219 m a.s.l. (see Figure 3). This is a well-
stratified, open-air site.

Physiographically, the area corresponds to the Southwestern
border of the Interserrana area, at the foothill of the

41

Figure 3. Archaeological sites cited in the text.

Ventania system.

Zoogeographically, the site is located in the Southern
District of Central Dominion (Ringuelet, 1961), where a
conjunction of species, predominantly, "guanaco", "pichi"
(**Zaedyus pichiy**) and skunk (**Conepatus sp.**), is observed.

Its climate indicates a mean annual temperature of 14° C and
a mean annual rainfall of 809 mm (see Figures 4 and 5). In
Thornthwaite's classification, the climate is $C_2 B_2'$ r a":
humid-subhumid, mesothermal with summer concentration of thermal
efficiency less than 48% (Rabassa, 1982).

Cañada de Rocha site is localized at 700 km northeast from
La Toma site, at lat. 34° 31' 05" S and long. 59° 09' W and
20 m a.s.l. (see Figure 3), within the Pampa Ondulada area. It
is also a well-stratified, open-air site.

Phytogeographically, the area belongs to the Eastern District in Pampean province (Cabrera, 1971), characterized by a gramineous seudosteppe, communities of rushes and thistles and small woods of carob trees.

Zoogeographically, the site is within the border between the Subtropical and Pampasic Dominion (Ringuelet, 1961) with a greater influence of the former. The more frequent mammals are "carpincho" (**Hydrochoerus hydrochaeris**), "nutria" (**Myocastor coypus**) and "ciervo de los pantanos" (**Blastocerus dichotomus**). Nowadays, this latter species appears farther north.

The climate indicates a mean annual temperature of 16.5° C; the mean annual rainfall is 950 mm (see Figures 4 and 5). According to Thornthwaite's classification, the climate is $B_1 B_2$ r a': humid, mesothermal, with low water deficiency and summer concentration of the thermal efficiency less than 48% (N. Porro, pers. com.).

DISCUSSION AND INTERPRETATION: The archaeological faunal sample in the palaeoenvironmental reconstruction and the palaeo-ecosystem exploitation

The faunal assemblages coming from archaeological sites represent a portion of the living fauna in the area when it was inhabited by the aborigines. Therefore, the fauna is just a valid variable so as to interpret the palaeoecosystem and the palaeozoogeography.

From this point of view, the Zooarchaeology goes beyond the archaeological knowledge, as it is of interest to the biological field. Likewise, the information from the Geological Sciences (Geomorphology, Sedimentology, Stratigraphy) in addition to the associated vertebrate, invertebrate, floral and human activity remains lead to formulate explicative hypotheses about the palaeoenvironment which the human groups were adapted to. Thus, the interdisciplinary work becomes very important.

Although the sites chosen have differences related to the length of the human occupation period (see Figures 6, 7 and 8), the detected records are useful to infer the cultural-ecological adjustment and, referring to the fauna, what species were characteristic in the area and which of them were exploited by indigenous people (see Figure 9).

Archaeological remains from Cave 3, in Los Toldos Locality (Figure 3; Cardich **et al.**, 1973) demonstrated that the human occupation extended from the Late Pleistocene until before the

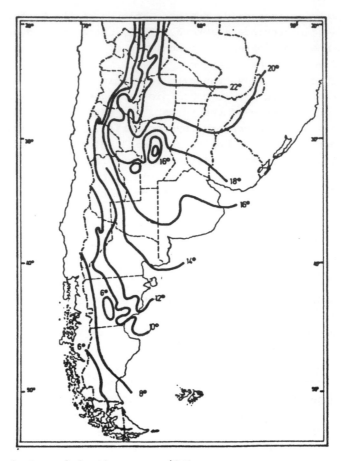

Figure 4. Annual isotherms map(°C).

Tehuelches adopted the horse in the XVIIIth century (see Figure 6).

   Through sedimentological information and correlation with pollen diagrams of Central and Southern Chile (Heusser, 1974; 1983; Heusser **et al.**, 1981), it could be inferred that the environmental conditions would have not presented remarkable differences in the region during prehispanic occupations. Some little oscillations would have produced more humidity in this region and they would temporarily coincide with existence of Middle Holocene hunters carrying Casapedrense lithic industry, dated at 7260±350 years BP (Cardich **et al.**, 1973; Cardich, 1977).
Likewise, some modifications in the faunal structure due

44

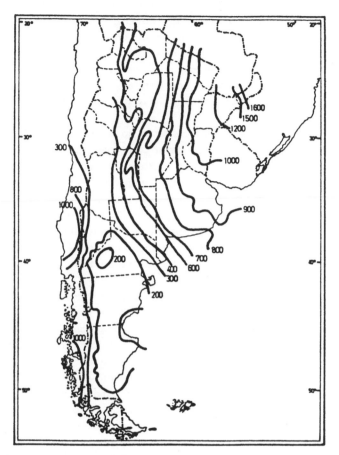

Figure 5. Average annual precipitation map(mm).

-at least in part- to these climatic variations were verified
through the vertebrates.

The most important changes detected in the faunal structure
of the bone sample are the extinct species, **Onohippion
(Parahipparion) saldiasi** and "**Lama gracilis**" (Cardich, 1984);
they were found associated with Level 11 Industry, which was
dated at 12,650±600 years BP and with Toldense Industry which
ended at 8750±480 years BP. Another species, **Rhea americana,**
was also found associated with the last mentioned industry.

The present habitat of this last species extends down south
to the Río Negro province (Tambussi and Tonni, 1985, and the
references cited there). Probably, the absence of this
flightless bird (**Rhea americana**) in the upper levels of the

45

sequence and the presence of another related bird (**Pterocnemia pennata**) are due to the fact that the ecological niche left by the former species when it searched for refuge to the north, was occupied by **P. pennata**, the species living in Los Toldos area at present (Cardich and Miotti, 1983; Tambussi and Tonni, 1985). This situation could indicate an environmental deterioration.

Nevertheless, from an economic point of view, the support of the aboriginal groups has been only one species (**L. guanicoe**) along the cultural sequence. Probably, the "guanaco" was the most used mammal by the three hunter groups that inhabited the cave at different times during the Late Pleistocene and Holocene. This could be due to the "guanaco" frequency in the area and/or because of adaptative adjustment of the human groups. This camelid inhabits the Patagonian and Central zoogeographical Dominions, being presently common in the area of the Santa Cruz plateau.

The two open-air sites selected from Pampean region, La Toma and Cañada de Rocha, were inhabited during the Late Holocene. There were, at least, two human occupations at La Toma site (see Figure 7). The faunal sample analized in this paper corresponds to the latest human settlement (995±65 yrs. BP). It is composed by species from Patagonian and Central Dominions (**L. guanicoe, Dolichotis patagonum, Zaedyus pichiy**) and the Pampasic Dominion (**Ozotoceros bezoarticus, Chaetophractus villosus**) whose association indicates arid or semiarid environments (Tonni, 1985). Only two species from Subtropical stock were recorded (**Dasypus hybridus** and **Cavia aperea**), although there were very few remains of them.

Considering that these last species have recently moved from the south of the Subtropical Dominion to the Pampean and Patagonian Dominions, it is formulated that the human occupation in La Toma occurred within arid or semiarid environments, which were characteristic conditions of a significant part of the Holocene (Tonni and Fidalgo, 1978; Fidalgo and Tonni, 1982). However, the presence of subtropical species indicates that the hunters who inhabited the site would have been in a transitional adaptive situation during which the environmental conditions would be changing from arid-to-semiarid cycles to more humid cycles, similar to the present one (Salemme et al., in press).

The other site, Cañada de Rocha, is located in Luján County (Buenos Aires province), in the floodplain of Luján River.

The archaeological remains have been deposited within

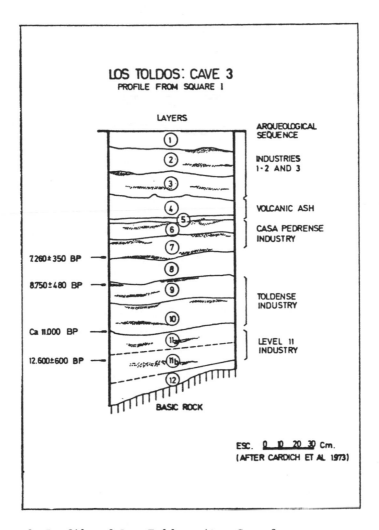

LOS TOLDOS: CAVE 3
PROFILE FROM SQUARE I

LAYERS

ARQUEOLOGICAL
SEQUENCE

INDUSTRIES
1-2 AND 3

VOLCANIC ASH

CASA PEDRENSE
INDUSTRY

7.260±350 BP

8.750±480 BP

TOLDENSE
INDUSTRY

Ca 11.000 BP

LEVEL 11
INDUSTRY

12.600±600 BP

BASIC ROCK

ESC. 0 10 20 30 Cm.
(AFTER CARDICH ET AL 1973)

Figure 6. Profile of Los Toldos site, Cave 3.

sediments assigned to the Holocene period. Because of the presence of ceramics as well as other cultural remains, this site has been referred to the Late Holocene (see Figure 8).

The faunal record (Ameghino, 1880) comprises species that are not presently found in the area and that are characteristic of the Patagonian and Central Dominions, for example **D.patagonum, L.guanicoe, Tolypeutes matacus,** Canis **(Pseudalopex) griseus, Felis concolor.** Besides, remains of **Lagostomus maximus** and **Ctenomys sp.** were recorded. These two species

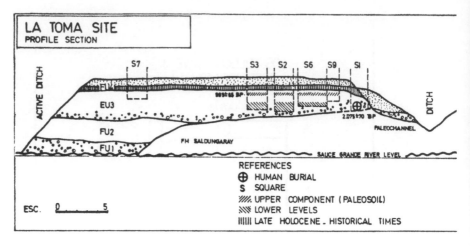

Figure 7. Profile section of La Toma site.

indicate local arid conditions and they are not found in the
site at present because of the recurrent, periodic floods
(Salemme, 1983).

Scarce remains corresponding to several species from the
subtropical stock have been found: **Blastocerus dichotomus**,
Hydrochoerus hydrochaeris and **Myocastor coypus**; these rodent
species indicate the presence of continental water bodies, but
their remains are not frequent in the site.

In agreement with this evidence, it has been formulated that
human occupation in Cañada de Rocha occurred under arid or
semiarid conditions (Salemme, 1983) which would have allowed
the displacement of central and patagonian species to the north
and to the east from their present habitat (Central or
Patagonian Dominions). Whereas, the species which presently
live in the area (or that could potentially be there) are
scarce or absent in Cañada de Rocha. In other sites of the
Pampa Ondulada (Río Luján site, Las Lechiguanas Island site,
Cañada Honda site), the more frequent species are those which
correspond to the Subtropical Dominion and indicate a faunal
composition similar to the present one. This fact suggests the
heterochrony of these sites with respect to Cañada de Rocha
as well as the human group adaptation in this last site during
an arid or semiarid cycle. These conditions would have been
very different from the present ones and from those under
which the other mentioned sites were inhabited.

FINAL CONSIDERATIONS

In summary, the partial reconstruction of the palaeoenvironments inhabited by aboriginal groups is drawn from the
examples exposed here. Besides, it is possible to formulate
some alternative hypotheses referred to the cultural-ecological
adaptations.

Although it must be taken into account that the length of the
settlements was not the same in the three cases, it can be
concluded that environmental modifications occurred during
their occupations, maybe due to climatic oscillations.

Los Toldos Cave 3 was inhabited since Late Pleistocene until
the Late Holocene, just before the historical times. The faunal
record leads to formulate an ecological deterioration near the
Late Pleistocene-Holocene boundary, which was evidenced by the
extinction of some pleistocenic megamammal species. Besides,
the presence of **Rhea americana** ca. 8700 years BP and its
absence in the record after this date would indicate an
environmental modification; certainly this species retreated
to northern Patagonia. Likewise, a more humid event close
to 7000 years BP could be inferred, based on sedimentological
evidence.

In spite of those environmental changes, **Lama guanicoe**,
which was the main animal resource for the aborigines,
predominates throughout the entire sequence. Though, this
species is still living in the area under arid-semiarid
environments; thus, it could indicate the predominance of those
conditions since Late Pleistocene until today.

The latest human settlement in La Toma occurred when the
ecological conditions were somewhat different from the present
ones. The archaeological faunal sample registered mainly
species which belong to the Pampean, Central and Patagonian
Dominions; they are still living in the area though in a
reduced number, but there is a larger record of subtropical
species nowadays. Therefore, very few remains of only two
subtropical species, **Dasypus hybridus** ("mulita") and **Cavia
aperea** ("cuis") were recorded in the archaeological sample.
Thus, it can be inferred that the environmental conditions
were changing to a more humid cycle, which allowed the
displacement of these species south of their subtropical
ecological niche.

Consequently, it can be formulated that the aborigines were
in a transitional adaptative situation to the new ecological
characteristics ca. 1000 years BP. However, the species mainly

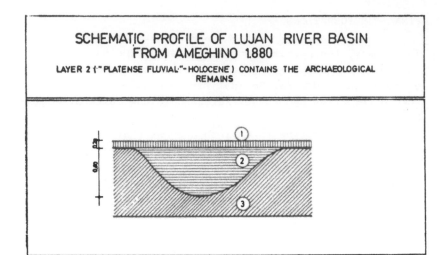

Figure 8. Schematic profile of Luján River Basin (from Ameghino, 1880). Layers 1,2 and 3: geological units. Layer 2: "Platense fluvial" - Holocene contains the archaeological remains. Thicknesses in meters.

exploited such as, "guanaco" and pampean deers as the major preys and the edentats, "piche' and "peludo", as complementary resources, came from the Pampean and Patagonian stocks.

The third studied case, Cañada de Rocha site, was inhabited under arid or semiarid climatic conditions during the Late Holocene. This is inferred from the faunal sample; it was widely composed of species from the central and patagonian stocks. Those species are not presently in the area, which is totally influenced by the Subtropical Dominion fauna.

Thus, Cañada de Rocha shows the existence of a prehispanic human occupation in an environment which was different from that one existing today. The ecological change is clearly evidenced by comparison of the present faunal list and the archaeological one.

Under those conditions, the aborigines were adapted to "guanaco" and deer hunting, which were the main indigenous economical resources.

ACKNOWLEDGEMENTS

Basic work for this investigation was performed through

TAXA	LOS TOLDOS			LA TOMA		CAÑADA DE ROCHA
SITES	Level 11 Toldense	Casapedrense	Level 1,2 and 3	Latest occupation	Earlier occupation	occupation
Gastropoda						x
Adelomedon brasiliensis				x		
Zidona duffresnei				x		
Polyborus sp.	x					
Rhea cf.R.americana	x			x		
Pterocnemia cf.P.pennata			x			
Eudromia elegans	x					
Canis (Pseudalopex) griseus						x
Canis (P.) gymnocercus						x
Canis (P.) sp.	x	x				
Canis (Canis) familiaris		x				
Chrysocyon brachyurus			–			x
Felis onca						x
Felis concolor	x			x	x	x
Conepatus chinga						x
Conepatus sp.			x			
Tolypeutes matacus						x
Zaedyus pichiy				x	x	
Chaetophractus villosus				x	x	x
Dasypus hybridus				x		x
Onohippidion (P.) saldiasi*	x	x				
Lama cf.L.gracilis*	X					
Lama guanicoe	x	x	x	x	x	x
Lama sp.					x	
Ozotoceros bezoarticus				x	x	x
Reithrodon auritus						x
Calomys laucha						x
Cricetidae gen.et sp.indet.	x	x				
Cavia aperea				x		x
Dolichotis patagonum				x		x
Caviidae gen.et sp.indet.		x				
Ctenomys sp.				x	x	x
Myocastor coypus						x
Lagostomus maximus						x

*Extinct species.

Figure 9. List of taxa registered in the studied sites.

research scholarships from CIC (Buenos Aires Province) and
CONICET.

The authors are greatly indebted to the following scientists,
who made interesting observations to the previous version of
the manuscript: Dr E. Tonni, Dr G. Politis and Dr E. Mansur-
Franchomme.

REFERENCES

Ameghino, F.: "La antigüedad del hombre en El Plata. Obras
    completas y correspondencia científica" Vol. III Gob. Pcia
    Bs. As., La Plata, 1880.
Aschero, C. "Interrogación sobre actividades en el sitio
    precerámico Inca-Cueva 4. Resúmenes de los Trabajos presen-
    tados al VIII Congreso Nacional de Arqueología Argentina".
    Concordia. 1985.
Bettinger, R.: "Explanatory Predictive models of hunter-
    gatherers adaptations". In: Advances in Archaeological
    Methods and Theory, vol. 3:189-255. Schiffer (ed.),
    Academic Press. 1978.
Binford, L.: "Bones: ancient men and modern myths". New York,
    Academic Press, 1981.
Borrero, L. "La relación entre los cazadores americanos y la
    fauna pleistocénica: consideraciones demográficas". In: Actas
    II Congr. Arg. Paleontología y Bioestratigrafía y I Congr.
    Latinoamericano de Paleontología. Vol. 3, Buenos Aires. 1978.
Borrero, L.; Casiraghi, M. & Yacobaccio, H. "First guanaco-
    processing site in Southern South America" Current
    Anthropologist, 26(2):273-276. 1985.
Bottema, S.: "The use of gastroliths in archaeology". In:
    Archaeozoological studies. Clason, A. (ed.), North Holland
    Publ. Amsterdam, 1975.
Browman, E. "Prehistoric pastoralism nomadian of the Jauja,
    Huancayo Basin central Perú". Annual meeting Society of
    American Archaeology, Florida, 1974.
Butzer, K. "Context in Archaeology: an alternative Perspective".
    Journal of field Archaeology, vol. 7 N° 4, Boston Univ.,
    Boston. 1980.
Cabrera, A. "Fitogeografía de la República Argentina" Bol.
    Soc. Arg. Bot. 14(1-2):1-42. Buenos Aires. 1971.
Cabrera, A. & Yepes, J.: "Mamíferos sudamericanos", Ediar (ed.),
    1:1-187; 2:1-160. Buenos Aires, 1960.

Cardich, A. "Las culturas pleistocénicas y post-pleistocénicas de Los Toldos (Santa Cruz, Argentina)." In: tomo Centenario del Museo de La Plata, tomo 1, Antropología 1977.

Cardich, A.; Cardich, L. & Hajduk, A. "Secuencia arqueológica y cronología radiocarbónica de la Cueva 3 de Los Toldos (Santa Cruz, Argentina)". In: Relaciones de la Soc. Arg. de Antrop. t. VII, n.s. 1973.

Cardich, A.: "Palecambientes y la más antigua presencia del hombre". In: Las culturas de América en la época del descubrimiento. Cultura Hispánica (ed.). Madrid: 1-36. 1984.

Cardich, A. & Laguens, A.: "Fractura intencional y posterior utilización del material óseo arqueológico de la Cueva 3 de Los Toldos, Pcia de Santa Cruz, Argentina". Rev. Museo La Plata (n.s.), T. VIII, Antropología 63. 1984.

Cardich, A. & Miotti, L.: "Recursos faunísticos en la economía de los cazadores-recolectores de Los Toldos (Pcia. de Santa Cruz, Argentina). Relaciones, Soc. Arg. Antrop., T. XV, n.s. 145-157. 1983.

Cigliano, E.: "Tastil, una ciudad preincaica argentina" Cabargón (ed.), Buenos Aires. 1973.

Clason, A. Archaeozoological Studies. North Holland Publ. Amsterdam, 1975.

Correal Urrego, G. Evidencias culturales y megafauna pleistocénica en Colombia. Fundación Invest. Arqueol. Nacionales, Banco de la República, Bogotá. 1981.

Crawford, R. "A través de Las Pampas y Los Andes" Eudeba, 228pp. 1976.

Darwin, Ch: "A naturalist's voyage round the world in H.M.S. Beagle (spanish version): "Un naturalista en el Plata" Centro Editor América Latina, 1977, Buenos Aires, 1846.

Daus, F.: "Morfografía general de las Llanuras argentinas". Geografía de la Rep. Arg., GAEA, II:115-195, Buenos Aires, 1946.

Davis, L.: "Climate changes and the advent of domestication: the succession of ruminant artiodactyls in the late Pleistocene-Holocene in the Israel Region" Paleorient. Vol. 8/2. 1982.

Falkner, T.: "Descripción de la Patagonia". Hachette (ed.), 1974. 174pp. Buenos Aires, 1910.

Fidalgo, F. ; Meo Guzman, L.; Politis, G.; Salemne, M. & Tonni, E. Investigaciones arqueológicas en el Sitio 2 de Arroyo Seco (Pdo. Tres Arroyos - Pcia de Buenos Aires - Rep. Argentina). New evidence for the Pleistocene Pecpling of the Americas. A. Bryan (ed.), Center for the Study of Early Man, Univ. of Maine. Orono. p.221-269. 1986.

Fidalgo, F. and Tonni, E.: "The Holocene in Argentina, South America". Chronost. Subdiv. of the Holocene, Striae, 16:49-52. Moscú. 1982.

Flannery, K.: The vertebrate fauna and hunting patterns. The Prehistory of the Tehuacán Valley, Vol. I, B. Meggers (ed.), Anthrop. Soc. of Washington, Washington D.C. 1967.

Foley, R.: Modelling hunting strategies and inferring predator behaviour from prey attributes. BAR series, Hunters & their preys: p.63-76, England, 1983.

Grant, A.: The use of tooth wear as a guide to the age of domestic ungulate. Ageing and sexing animal bones from archaeological sites BAR series, 109. Wilson, Grigson & Payne (eds.) England, 1982.

Gusinde, M.: Die Selkhmanm. Mödling bei Wien. Verlag der Internationalen zeitsch ifl Anthropos., Die Feuerland Indianer, 3 vol. 1931.

Heusser, C.: Vegetation and climate of the Southern Chilean Lake District during and since the last Interglaciation. Quaternary Research, 4:290-315.

Heusser, C.: "Quaternary palinology of Chile". In: Quaternary of South America and Antarctic Peninsula, 1(2):5-22, Rotterdam. 1983.

Heusser, C.; Streeter, S. & Struiver, M.: "Temperature and precipitation in Southern Chile extended to 43.000 yr ago". Nature, 294:65-67. 1981.

Jochim, M.: "Hunter-gatherer subsistence and settlement: A predictive model". Academic Press, New York. 1976.

Klein, R. & Cruz-Uribe, K.: "The analysis of animal bones f from Archaeological sites". Univ. of Chicago Press, 249. 1984.

Melendez, M.: "Paleontología". In: Paraninfo (ed.). Madrid.

Mengoni Goñalons, G.: Prehistoric utilization of faunal resources in arid Argentina. Animals and Archaeology: 1. Hunters and their Prey. J. Clutton Brock & C. Grigson (eds.) BAR series. 163:325-335. England. 1983.

Moreno, F.: Viaje a la Patagonia Austral. Solar/Hachette (eds.) 1969.

Morlan, R.: Taphonomy and archaeology in the Upper Pleistocene of the northern Yukon territory: a glimpse of the peopling of the New World. Paper N° 94 Arch. Survey of Canada, Ottawa. 1980.

Musters, G. Vida entre los Patagones, Biblioteca Centenaria, Univ. Nac. La Plata, 1:127-338. La Plata, 1911.

Olson, S.: Zooarchaeology: animal bones in archaeology and their interpretation. Addison-Wesley Reading. 1971.

Olson, S.: Biological archaeology in the West Indies. The Florida Anthropologist. Vol. 35 N°4:162-168. 1982.

Outes, F.: Un texto Aonükün'k (Patagón meridional). Rev. Mus. La Plata, T. XXXI:353-369. La Plata. 1928.

Oviedo y Valdez, G.: Historia general y natural de las Indias islas de Tierra firme del Mar Océano, José Amador de los Ríos (ed.), 4Vol. Madrid. 1851.

Payne, S.: On the interpretation of bone samples from archaeological sites. Papers in Economic Prehistory. Higgs, E. (ed.), Cambridge Univ. Press. 1972.

Palanca, F. & Politis, G. Los cazadores de fauna extinguida de la Provincia de Buenos Aires. Prehistoria Bonaerense: 70-91. Olavarría. 1979.

Parodi, L.: "Fitogeografía de la Argentina". In. Geog. Rep. Arg. GAEA VIII. Buenos Aires. 1947.

Pigafetta, A.: Il primo viaggio in torno al Globo e le sue Régole sull' Arte del Navigare. Raccolta di Documenti e Studi pubblicati Dalla Commissione Colombiana pel 4° Cent. Scoperta dell' America, parte V, vol. II y III. Roma 1894.

Politis, G.: Arqueología del área Interserrana bonaerense. Tesis Doctoral. Fac. Cs. Nat. y Museo Universidad La Plata 392. 1984.

Politis, G.: Recientes investigaciones en "La Moderna". Un sitio de matanza de **Doedicurus clavicaudatus** en la Región Pampeana (Argentina). Resumen 45° Cong. Internac. Americanistas 393, Bogotá, 1985.

Politis, G. & Tonni, E. Arqueología de la Región Pampeana: el sitio 2 de Zanjón Seco, Partido de Necochea, Prov. de Buenos Aires, Rep. Argentina. Rev. Pre-historia, Univ. de Sao Paulo, Inst. de Pre-Historia, vol. III(4):107-139, 1983.

Pollard, G. Identification of domestic **Lama sp** from prehispanic northern Chile using microscopy. El Dorado Newsletter Bull. South America Anthrop. 1976.

Priegue, C.: "La información etnográfica de los Patagones del siglo XVIII". Cuadernos del Sur. 3. Univ. Nac. del Sur. Bahía Blanca. 1971.

Rabassa, J.: "Variación regional y significado geomorfológico de la densidad de drenaje en la cuenca del Río Sauce Grande

Provincia de Buenos Aires". Rev. Asoc. Geol. Arg., XXXVII (3): 268-284. 1982.

Raffino, R.; Tonni, E. & Cione, A.: "Recursos alimentarios y economía en la región de la Quebrada del Toro, Provincia de Salta, Argentina. Relaciones. Soc. Arg. Antrop. XI (NS) 9-30. Buenos Aires. 1977.

Ringuelet, R.: "Rasgos fundamentales de la Zoogecgrafía de la Argentina". Physis 22 (63) 151-170. Buenos Aires.

Salemme, M.: Distribución de algunas especies de mamíferos en el Noreste de la provincia de Buenos Aires durante el Holoceno. Ameghiniana, XX(1-2):81-94.

Salemme, M. & Tonri, E.. Paleoetnozoología de un sitio arqueológico en la Pampa Ondulada: Sitio Río Luján (Partido de Campana, Prov. de Buenos Aires). Relaciones, Soc. Arg. Antrop., XV(n.s.):77-90. 1983.

Salemme, M.; Politis, G.; Madrid, P.; Oliva, F. & Güerci, L. Informe preliminar sobre las investigaciones arqueológicas en el Sitio La Toma, partido de Cnel. Pringles, prov. de Buenos Aires. VIII Cong. Nac. Arg. Concordia. 1985, 18 al 22 de mayo.

Service, M. Los cazadores, Nueva Col. Labor (ed.), España, 1966.

Shimada, M.: Zooarchaeology of Huacaloma: behavioural and cultural implications. Appendix IV:303-336. In: Investigations at Huacaloma, Shimada (eds.).

Silveira, M.: Análisis e interpretación de los restos faunísticos de la Cueva Grande del Arroyo Feo. Relaciones Soc. Arg. Antrop., XIII (n.s.):229-253. 1979.

Sivertsen, B.: A site activity model for kill and butchering activities at hunter-gatherer sites. Jour. of Field Archaeol. Vol. 7(4):423-442. 1980.

Smith, P.: Toward a more accurate estimation of the meat yield of animal species at archaeological sites. Archaeol. Studies, A. Clason (ed.), North Holland Publ., Amsterdam. 1975.

Speth, J.: Bison kills and bone counts. Prehist. Archaeol. & Eccl. series. Butzer & Freeman (eds.), Univ, Chicago Press. 1983.

Tambussi, C. & Tonni, E.: Aves del Sitio Arqueológico Los Toldos, Cañadón de las Cuevas, Provincia de Santa Cruz (República Argentina). Ameghiniana, 22(1-2):69-74. 1985.

Tonni, E.: La Arqueología Biológica en la Argentina: el estudio de los vertebrados. Adeha, VI:3-11. Buenos Aires, 1984.

Tonni, E.: Mamíferos del Holoceno del Partido de Lobería,

Provincia de Buenos Aires. Aspectos paleoambientales y bio-
estratigráficos del Holoceno del sector oriental de Tandilia
y Area Interserrana. Ameghiniana, 22(3-4):283-288. 1985.

Tonni, E. & Fidalgo, F. Consideraciones sobre los cambios
climáticos durante el Pleistoceno Tardío-Reciente en la
Provincia de Buenos Aires. Aspectos ecológicos y zoogeográficos relacionados. Ameghiniana, 15(1-2):235-253. 1978.

Wing, E.: Utilization of animal resources in the Peruvian Andes. Appendix 4. Excavation at Kotosh, Perú, 1963 and 1966.
Izumi y Terada (eds.) p.327-351. Univ. Tokyo, 1972.

Wing, E.: American domestication in the Andes. Origins of
Agriculture. C.E. Reed (ed.). Mouton Press, Hague, 1977.

Wood, R. & Johnson, D.: A survey of disturbance processes in
archaeological site formation. In: Adv. Mth. and Theory,
vol.I, M. Schiffer (ed.), Academic Press, 1978.

MIGUEL ALBERO & FERNANDO E. ANGIOLINI

4

Instituto de Geocronología y Geología Isotópica (INGEIS), Buenos Aires, Argentina

ERNESTO L. PIANA

Centro Austral de Investigaciones Científicas, Ushuaia, Tierra del Fuego, CONICET, Argentina

Holocene ^{14}C reservoir effect at Beagle Channel (Tierra del Fuego, Argentina Republic)

ABSTRACT

The evaluation of the "reservoir effect" as a function of time in Beagle Channel is presented in this paper. Contemporaneous charcoal and **Mytilus edulis** shells sampled in sites from three different localities along the channel were ^{14}C dated and their age differences were calculated. Charcoal ages, considered as "true ages", range from 5600 to 360 ^{14}C yr B.P. and age differences between charcoal and **Mytilus** show good reproducibility except for one of the localities. Marine and terrestrial samples from such locality show smaller differences which are attributed to the local influence of a river with a high organic matter content.

An age difference of 556 $\pm$ 61 years is proposed to correct ^{14}C dates of marine organisms in this area.

RESUMEN

En este trabajo se presenta una evaluación del "efecto reservorio" en función del tiempo para el canal de Beagle. Para esta evaluación se han datado por ^{14}C muestras de carbón y conchas de **Mytilus edulis** provenientes de sitios de tres localidades próximas al canal y se han calculado las diferencias de edad entre carbón y **Mytilus** muestran una buena reproducibilidad si no se considera una de las localidades muestreadas. Las muestras de la localidad exceptuada exhiben diferencias menores que se atribuyen a la influencia local de un río con alto contenido de materia orgánica.

Se propone la cifra de 556 $\pm$ 61 años como corrección para las edades ^{14}C de organismos marinos en ese área.

INTRODUCTION

The depletion of the $^{14}C/^{12}C$ ratios of marine organisms in comparison with those of continental plants and animals causes ^{14}C ages of marine shells to be some hundreds of years greater than real ages, since their ^{14}C concentration reflects the isotope content of the water in which the dated organisms lived.

The magnitude of this depletion known as "reservoir effect" is a funtion of geographic location and, eventually, of time as a consequence of the different origins of the relevant local water masses. Therefore, for the correction of radiocarbon dates of marine specimens, the apparent age of the water masses in which the dated organisms lived must be estimated.

The particular geographic location of the Tierra del Fuego with three important connection ways between archipelago of the Pacific and the Atlantic oceans (Strait of Magellan, Beagle Channel and Drake Passage), makes especially attractive the study of the reservoir effect, in connection with sea-water dynamics.

A recent work (Rabassa **et al**, 1986) presents the first reliable date on the recent geologic history of the Beagle Channel area. Following these authors, "it seems very likely that the present Beagle Channel was occupied by a glacial lake after deglaciation. This lake was later replaced by sea water, following the opening of the channel perhaps sometime before 8200 yr B.P. The marine environment was fully established along the channel and perhaps surrounding the Isla Grande around 7900 yr B.P."

Such a short Holocene history as a maritime way encourage us to enlarge the study of the reservoir effect (as an indicator of sea water dynamics changes) throughout a long period of marine process and evolution. It is very likely that this aim may be achieved due to :

(i) numerous archaeological sites along the Beagle Channel shores, which prospections have been carried out by one of us (ELP), indicates that association of marine organism remains with charcoal is present in most of these sites and contemporaneity of these materials can be assessed (sensu Albero **et al,** 1986)

(ii) presence of human population at Tierra del Fuego at least during the last 10,000 years (Massone, 1983; Orquera and Piana 1986)

(iii) oldest presence of human remains taking advantage of maritime resources though scarcely - were found along the Beagle Channel shores dates ca 7000 yr B.P. (Orquera and Piana 1986). As it has been already mentioned the reservoir effect depends on geographic location and time, and our knowledge of its magnitude is obtained through the comparison of radiocarbon activity between contemporaneous marine and terrestrial organisms.

In a previous contribution (Albero **et al**, 1986) ^{14}C ages on different materials corresponding to a single archaeological

Table 1: Dated materials, $\delta^{13}C$ values and ^{14}C ages normalized to $\delta^{13}C$ - 25.0 % from Túnel I, D level (after Albero **et al**, 1986)

Sample Mo	Material	Species	δ^{13} C(‰)	C age (yr B.P.)
AC 0683	Charcoal		- 24.0	5630 ± 120
0693	Shell		3.4	6220 ± 120
0697	Shell		2.8	6340 ± 130
0695	Shell		3.0	5110 ± 130
0694	Shell		2.1	6180 ± 120
0696	Shell		0.0*	6290 ± 120
0676	Shell		4.9	5720 ± 120
0703	Bone Collagen		- 20.5	5280 ± 100
0698	Bone collagen		- 11.8	6220 ± 140

* Estimated value

event were presented (Table 1). In that paper, marine samples showed relative small age dispersion, if we do not take into account **Trophon** and **Balanus** sp samples.

The age coincidence between shells of **Nacella, Fisurella, Mytilus, Aulacomyas** shells and bones of **Arctocephalus australis** bones enables us, up to the moment, to consider any of these species as reliable. The arguments so as to consider charcoal age as "true age" have been discussed too (Albero **et al,** 1986).

In the present paper, attention will be paid to the dependence of the magnitude of the reservoir effect on time and location at the Beagle Channel through ^{14}C data of charcoal and **Mytilus edulis** archaeological remains from different sites and ages.

METHODOLOGY AND RESULTS

More than a decade of archaeological field work in southernmost shell-middens like Túnel I, Túnel II, Lancha Packewaia, Shumakush I and Shumakush X lead to the development of excavation and sampling techniques specially outlined to distinguish short term anthropogenic shell refuse accumulations (Orquera and Piana 1985; Piana, 1984).

Along the Beagle Channel shores a large number of such anthropogenic shell middens have been found in which charcoal and marine fauna are associated. From the latter, the most common specie is **Mytilus edulis**, which has been one of the reliable species selected in the previous work (Albero **et al,**

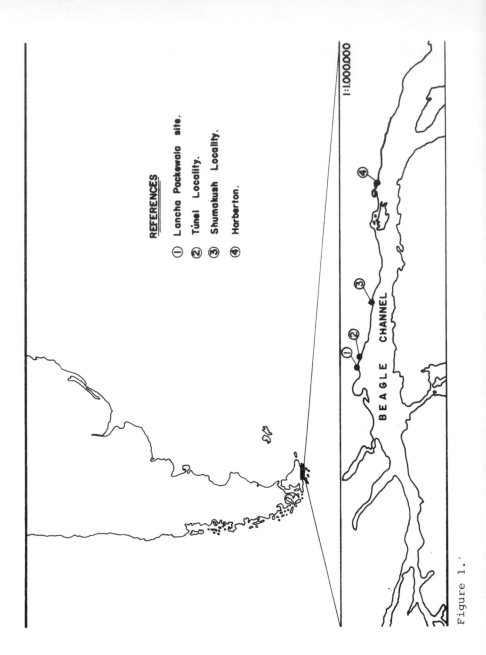

REFERENCES

① Lancha Packewaia site.
② Túnel Locality.
③ Shumakush Locality.
④ Harberton.

1:1.000.000

BEAGLE CHANNEL

Figure 1.'

1986) and due to its high relative frequency, it was chosen as representative to obtain the apparent age of marine organisms.

Samples from three different localities were collected to study the temporal and spatial behaviour of the reservoir effect (Fig. 1). Those from Túnel I, Túnel II, Shumakush I and Shumakush X were recovered during the 1984-85 and 1985-86 archaeological field works (Orquera and Piana, 1985; Piana and Orquera, 1985). Those from Shumakush II, III, V and IX and Harberton were sampled during 1984 and 1985 prospections by one of us (ELP)./

Carbon 14 dates were performed at INGEIS ^{14}C laboratory using liquid scintillation counting of synthesized benzene (Albero and Angiolini, 1983). Pretreatment of shells comprised hot H_2O_2 treatment plus 2 % HCl outer layer removal (ca 15 % loss in weight).In spite of the negligible humus content of the midden, charcoal was pretreated following laboratory routine, which consists of several washings with 2 % HONa at 100°C for 30 minutes until a clear extract is obtained followed by 2% HCl washings (ca 80°C) until all the carbonate is removed. This procedure has been proved to yield pure samples without any contamination.

δ ^{13}C measurements were performed at INGEIS Stable Isotopes Laboratory using a double-collector mass spectrometer Micromass 602-D (Panarello et al, 1981). Error in δ[13] C measurements is ± 0.3‰. All ^{14}C ages and Δ values presented in this paper are normalized to δ [13]C = - 25.0‰(Stuiver and Pollach, 1977). Samples δ ^{13}C were estimated at ‰ for shellsand - 25‰ for charcoal when measured values were not available.

Location of the archaeological sites is shown in Figure 1.

In order to evaluate the temporal evolution of the reservoir effect at Túnel site, the previously obtained data from "D" shell midden were compared with those from α shell midden, and completed with Túnel II - B shell midden covering from some 6000 to 1100 yr B.P. Results are presented in Table 2.

These results seem to be consistent with the difference of 765 ± 320 years found at Lancha Packewaia between charcoal (MC- 1068; 4215 ± 305 yr B.P.) and **Arctocephalus australis** bones (CSIC-307; 4980 ± 70 yr B.P.) (Orquera et al,1978). This archaeological site is located about 1km west from Túnel (Fig.1). It must be pointed out that charcoal sample was scarce and results come from two different laboratories.

Additional samples were collected at Estancia Harberton and Estancia Remolino. Results from Harberton (Table 3) show age differences of 530 ± 140 and 440 ± 130 years, which are consistent with those already presented in Table 2.

These results suggest that there was no significant variation in the reservoir effect in the considered period between 5630 and 360 years.

To complete this scheme, samples from different sites in Estancia Remolino, covering the last 1200 radiocarbon years, were taken. These sites are named according to the aboriginal.

Table 2. Charcoal and Mytilus ages from Túnel I and Túnel II sites

Archaelogic Site	Midden	Charcoal Sample Age and δ^{13}C	Mytilus Sample Age and δ^{13}C	Age difference (years)	Δ ‰
Túnel I	D	AC-0683 5630 ± 120 yr B.P. δ^{13}C=-24.0 ‰	AC-0694 6180 ± 120 yr B.P. δ^{13}C=-2.1 ‰	550 ± 170	-66 ± 20
Túnel I	α	AC-0677 3030 ± 100 yr B.P. δ^{13}C=-23.5 ‰	AC-0686 3780 ± 110 yr B.P. δ^{13}C=2.1 ‰ AC-0687 3700 ± 110 yr B.P. δ^{13}C=2.2 ‰	710 ± 130	-84 ± 14
Túnel II	B	AC-0824 1120 ± 90 yr B.P. δ^{13}C=-23.6 ‰	AC-1046 1670 ± 90 yr B.P. δ^{13}C=-25.0 ‰ *	550 ± 130	-66 ± 14

* estimated value

Table 3. Charcoal and Mytilus ages from Estancia Harberton sites

Archaelogic Site	Midden	Charcoal Sample, Age and $\delta^{13}C$	Mytilus Sample Age and $\delta^{13}C$	Age difference (years)	Δ (‰)
Harberton Galpón	Sup	AC-1040 1210 ± 90 yr B.P. $\delta^{13}C=-25.0$ ‰*	AC-1042 1740 ± 100 yr B.P. $\delta^{13}C=2.2$ ‰	530 ± 140	-64 ± 15
Harberton Cementerio	Sup	AC-1043 360 ± 90 yr B.P. $\delta^{13}C=-24.0$ ‰	AC-1045 800 ± 90 yr B.P. $\delta^{13}C=2.3$ ‰	440 ± 130	-54 ± 14

* Estimated value

Table 4. Charcoal and Mytilus ages from Estancia Remolino sites

Archaelogic Site	Midden	Charcoal Sample Age and $\delta^{13}C$	mytilus Sample Age and $\delta^{13}C$	Age difference (years)	Δ (‰)
Shumakush I	D sup	AC-0827 1220 ± 110 yr B.P. $\delta^{13}C=-25.2$ ‰	AC-0878 1280 ± 100 yr B.P. $\delta^{13}C=2.5$ ‰	60 ± 150	- 8 ± 18
Shumakush I	D 290	AC-1047 940 ± 110 yr B.P. $\delta^{13}C=-25.0$ ‰ *	AC-1033 1320 ± 90 yr B.P. $\delta^{13}C= 1.9$ ‰	380 ± 140	-46 ± 16
Shumakush II	Sup	AC-0826 1100 ± 120 yr B.P. $\delta^{13}C=-24.8$ ‰	AC-0877 1320 ± 100 yr B.P. $\delta^{13}C=2.4$ ‰	220 ± 160	-28 ± 18
Shumakush IV	Sup	AC-0829 310 ± 100 yr B.P. $\delta^{13}C=-25.0$ ‰ *	AC-0880 450 ± 100 yr B.P. $\delta^{13}C=1.9$ ‰	140 ± 140	-21 ± 17

Shumakush III Sup	AC-0828 410 $\pm$ 100 yr B.P. $\delta^{13}C$=-23.9 %₀	AC-0879 700 $\pm$ 100 yr B.P. $\delta^{13}C$= 2.6 %	290 $\pm$ 140	-36 $\pm$ 16
Shumakush IX Sup	AC-0830 940 $\pm$ 100 yr B.P. $\delta^{13}C$=-23.9 %₀	AC-0881 960 $\pm$ 100 yr B.P. $\delta^{13}C$= 2.4 %₀	20 $\pm$ 140	- 3 $\pm$ 17
Shumakush X Sup	AC-0832 440 $\pm$ 100 yr B.P. $\delta^{13}C$=-24.9 %₀	AC-0882 690 $\pm$ 110 yr B.P. $\delta^{13}C$= 3.6 %₀	250 $\pm$ 150	-30 $\pm$ 17

*Estimated value

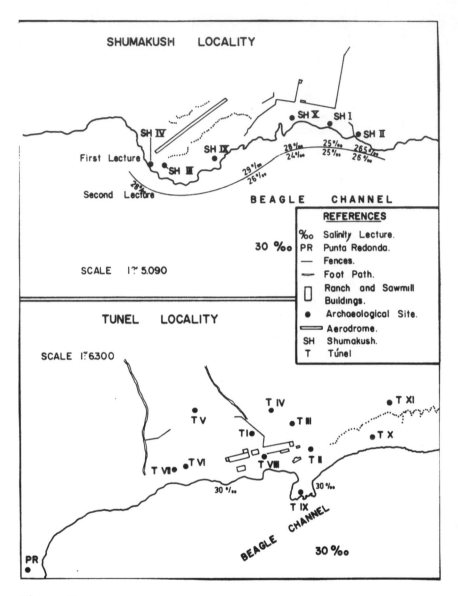

Figure 2.

These results clearly do not match with the previous sets, because the amplitude of the age differences between **Mytilus** and charcoal are significantly smaller than the previously reported in Tables 2 and 3. The discordance encouraged us to analyze the following hypoteses:
i) a variation of the reservoir effect during last millenium

ii) particular environmental conditions at this locality that produced a reservoir effect which is different from the others registered in the area

Alternative (i) had be rejected because samples of comparable ages in other localities (Túnel II and Estancia Harberton vs Shumakush I, II and IX) have yielded a greater reservoir effect than the ones discussed above. Therefore alternative (ii) had to be regarded.

The major difference between this area with distinct values when compared with Túnel and Harberton is the presence of freshwater influence of Remolino river. Estimated media discharge of this stream is only about 0.5 m^3/seg (data provided by the Programa de Hidrometeorología - CADIC); in such conditions, freshwater influence is evidenced by a local depletion of salinity up to 6‰ with respect to Beagle Channel sea-waters that show a media of 30‰ (data by Dr A.Amor, CONICET, using a hand refractometer, Figure 2). Highest discharge peaks, during the thaw season, reach about 5m^3 /seg. Río Remolino basin catchment extends for some 30 sq.km and is almost covered by a dense **Nothofagus** forest and receives some water from modern peat bogs as well. All this situation results in a concentration of suspended organic matter and, consequently, of dissolved carbonates.

This particular scenario is suspected to produce higher ^{14}C concentrations at the Channel in the neighbourhood of the stream mouth, causing that the reservoir effect is diminished in the area.

CONCLUSIONS

When marine organisms are ^{14}C dated, careful attention should be paid to influence of reservoir effect.

Not all marine shells have proved to be reliable and a different behaviour is shown by different species. According to our present knowledge, only **Mytilus** , **Aulacomya, Nacella** and **Fisurella** seem to be reliable so as to obtain the magnitude of the reservoir effect by comparison of their ^{14}C ages with those of charcoal, at least at the Beagle Channel area (Albero et al, 1986). Local conditions should be regarded as well: i.e. freshwater availability may modify ^{14}C apparent ages of marine organisms with low or non-mobility.

If the data obtained for Estancia Remolino are not taken into account, based on the reasons mentioned above, data obtained from Tunel and Harberton suggest that the reservoir effect has been almost constant in the period between 5600 and 360 yr B.P., though the available data do not imply an absolutely conclusive evidence of such a constancy, for the entire period.

The 5 pairs of samples measured give an average of 556 ± 61 years or Δ = (-67 ± 7) ‰ that is consistent with the previously reported data (Albero **et al**, 1986), with the previously

mentioned information from Lancha Packewaia and with other
reported values from the Northern Hemisphere (Olsson, 1980).

We therefore suggest that ^{14}C dates of those marine shells
considered as being reliable for the Beagle Channel should be
corrected by sustraction of this apparent age. A higher degree
of confidence will probably emerge from series of data obtained
at places where present or ancient fresh-water influence may be
discarded.

ACKNOWLEDGMENTS

We thank L.A. Orquera for the critical reading of the manuscript
and E. Linares for his support. We are also grateful to C.
Schroeder and R. Iturraspe for measurements of discharge and
basin catchment of Río Remolino river and A.Amor for salinity
determinations. We are also indebted to J.L.Nogueira and E.
Schluter of the ^{14}C Laboratory of INGEIS, and our colleagues
of the Stable Isotopes Laboratory of INGEIS, Servicio de
Cartografía of CADIC and the Asociación de Investigactiones
Antropológicas. This work is contribution N°99 of INGEIS.

REFERENCES

Albero, MC and Angiolini, FE, 1983, INGEIS radiocarbon
  laboratory dates I: Radiocarbon, v 25, no 3, 831-842
Albero, MC, Angiolini, FE and Piana EL, 1986, Discordant ages
  related to reservoir effect of associated archaeologic
  remains from the Túnel Site, Beagle Channel, Argentine
  Republic: Radiocarbon, v 28, no. 2A, 748-753
Massone, M: 1983, 10 400 años de Colonización Humana en Tierra
  del Fuego. In. Rev. Infórmese Año II N°14, Punta Arenas
  24-32
Olsson, IU, 1980, Content of ^{14}C in marine mammals from northern
  Europe. in Stuiver, M. and Kra, RS, eds. International
  Conf, 10th. Proc: Radiocarbon, v 22, no.3, 662-675
Orquera, LA, Piana, EL, Tapia, AH, 1978, Lancha Packewaia:
  Arqueología de los canales fueguinos. Ed Huemul, Buenos Aires
  259
Orquera, LA, Piana, EL, Tapia, AH 1984. Evolución adaptativa
  humana en la región del Canal de Beagle: presented at "Las
  Jornadas de Arqueología de Patagonia, Trelew, Argentina.
Orquera, LA; Piana, EL, Tapia, AH, 1985, Sexta, séptima y
  octava campañas arqueológicas en Tierra del Fuego: la
  localidad Túnel: presented at the VIII Congreso Nacional de
  Arqueología Argentina, Concordia
Orquera, LA and Piana EL, 1983, Prehistoric maritime adaptations
  at the Magellan Fuegian littoral: presented at New World
  Maritime Adaptation symposium, 48th ann Mtg Soc Am Archaeol
  Pittsburg
Orquera, LA and Piana EL 1986. Littoral human adaptation at the

70

Beagle Channel maximum possible age. In this volume

Panarello, HO, García, CM, Valencio, SA and Linares E, 1981,
determinación de la composición isotópica del carbono en
carbonatos. Su utilización en hidrogeología y geología.
Revista de la Asociación Geológica Argentina, v 35, no.4,
460-466

Piana, EL, 1984, arrinconamiento o adpatación en Tierra del
Fuego: Ensayos de Antropología Argentina, Ed. Univ de Belgrano
Buenos Aires, 5-110

Piana, EL and Orquera, LA, 1985, octava campaña arqueológica en
Tierra del Fuego: la localidad Shumakush: presented at the VIII
Congreso Nacional de Arqueología Argentina, Concordia

Rabassa,J, Heusser, C and Stuckenrath, R, 1986, New data on
Holocene sea transgression in the Beagle Channel, Tierra del
Fuego: International Symposium on sea-level changes and
Quaternary shore lines, Sao Paulo, Brazil, Julio 1986

Stuiver, M and Polach, HA, 1977, Reporting of ^{14}C data:
Radiocarbon, v 19, 355-363.

ALBERTO L.CIONE
División Paleontología Vertebrados, Museo de La Plata, Argentina, and CONICET

ALDO E.TORNO†
Museo Argentino de Ciencias Naturales 'B. Rivadavia', Capital Federal, Argentina

5

Records of *Pogonias cromis* (Perciformes, Sciaenidae) in Las Escobas Fm (Holocene) in Uruguay and Argentina – Zoogeographical and environmental considerations

ABSTRACT

Records of black drum (**P. cromis**) in sediments of the marine postglacial transgression near the present Río de la Plata estuary are listed and commented and their zoogeographical and environmental significance is discussed. The hypothesis proposed in this paper is that salinity in localities 1, 2, 3, 4 and 6 was at least similar to that presently recorded at Punta Piedras and the mouth of Río Santa Lucía or even, more marine.

RESUMEN

En este trabajo se detallan hallazgos de corvinas negras (**P. cromis**) en sedimentos de la transgresión marina postglacial en zonas cercanas al actual Río de la Plata, como así también se discuten aspectos zoogeográficos y ecológicos, proponiéndose la hipótesis de que la salinidad en las localidades 1, 2, 3, 4 y 6 era entonces por lo menos similar a la actualmente presente en Punta Piedras y en la boca del Río Santa Lucía o aún, más marina.

INTRODUCTION

Remains of black drums (**Pogonias cromis** (Linné, 1776) Perciformes, Sciaenidae) are very common in Las Escobas Formation (Holocene) in Buenos Aires Province, Argentina. They are also found in contemporary units in Uruguay. Considering that published records are scarce (Tonni and Cione, 1984; Perea and Ubilla, 1981), it seems necessary to provide an up-to-date record of these fishes in the Holocene of this region as well as to discuss its palaeoecological significance.

P. cromis belongs to Sciaenidae, a mainly marine family of cosmopolitan distribution in warm or warm-temperature waters. Sciaenids occur in units older than the Holocene in Argentina ("Belgranense", Pascua Fm, Pleistocene, Ameghino, 1898; Paraná Fm, Upper Miocene, Cione and Torno, 1984; MS).

Las Escobas Fm corresponds to a post-glacial sea-level which was higher than the present one. Their sediments are found even tens of kilometers away from the recent marine coast, reaching a topographic level of approximately five meters a.m. s.l. They partially correspond to the "Platense marino"of Frenguelli. The unit overlays the Puesto Callejón Viejo Soil, the Guerrero Member of the Luján Fm or the "Pampeano" Fm. Radiocarbon dating yielded ages between 3000 and 6500 yr BP (Fidalgo and Martínez, in Tonni and Cione, 1984; Tonni and Fidalgo, 1978).

PALAEONTOLOGICAL RECORD: ARGENTINA (LAS ESCOBAS FM)

Locality: 100 m SW from the General B. Mitre Railway bridge
   on the Cañada Honda, Baradero County, Buenos Aires Province.
Material: pharyngeal teeth.
Repository: Museo Municipal de Ciencias Naturales "Carlos
   Ameghino", Mercedes, Buenos Aires Province.
Reference: unpublished.
Associated marine fishes: **Eugomphodus taurus**, cf **Myliobatis** sp.

Locality: 2 km south from the cross-road between the "Camino
   de Cintura" and General Ricchieri Highway, near Ezeiza
   Airport, in the limit of E. Echeverría and La Matanza
   Counties, Buenos Aires Province. The site is located 16 km
   southwest from Plaza Congreso in Buenos Aires City and 17 km
   from the present coast of the Río de la Plata.
Material: MLP 82-IV-5-1. Caudal fragment of skull.
Comments: Supraoccipital and left epiotic, parietal, pterotic
   fragments and the caudal part of the frontals are preserved.
   Exoccipital condyles and the caudal part of parasphenoides
   are also present (Figure 1). The material is assigned to
   P. cromis by the following morphological coincidences:
   a) Supraoccipital: The upper part of the crest is notably
   robust.

74

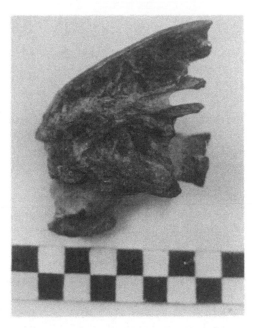

Figure 1. **Pogonias cromis**. Left lateral view of caudal portion
of neurocraneum. MLP 82-IV-5-1. Each square: 1 cm.

b) Parietal: The longitudinal ridge is mesially shifted,
being on the parieto-supraoccipital suture.
c) Epiotic: It has a very typical shape, caudally developed.
d) Pterotic: Similar robustness and position.
Repository: División Paleontología Vertebrados, Museo de La
Plata, La Plata, Buenos Aires Province.
Reference: unpublished.

Locality: excavation in the block limited by 122nd, 123rd, 43rd
and 44th streets, Ensenada County, Buenos Aires Province.
Canal 18 Member.
Material: 156 pharyngeal teeth, MLP 83-XI-10-49 to 205.
Repository: División Paleontología Vertebrados, Museo de La
Plata, Buenos Aires Province.
Reference: Tonni and Cione (1984).
Associated marine fishes: **Galeorhinus** cf. **vitaminicus**, cf.
Carcharhinus sp, cf. **Myliobatis** sp, Rajiformes indet.

Locality: Rinquelet, La Plata County, Buenos Aires Province.
Material: one pharyngeal tooth, MLP 85-I-20-1.

Comments: isolated tooth in the Frenguelli collection, without
further precision about location.
Repository: División Paleontología Vertebrados, Museo de La
Plata, La Plata, Buenos Aires Province.
Reference: unpublished.

Locality: left margin of La Ballenera creek, Boulevard Maríti-
mo (Mar del Sud), General Alvarado County, Buenos Aires
Province.
Material: one opercular, several vertebrae, dorsal rays, ribs,
N°6534.
Comments: isolated diagnostic bones in ancient collections.
Repository: Sección Paleontología Vertebrados, Museo Argentino
de Ciencias Naturales, Buenos Aires.
Reference: unpublished.

Locality: Cantera Bagnatti, Puerto Ruiz, Gualeguay Department,
Entre Ríos Province.
Material: one ray X of dorsal fin and one ray II of anal fin.
These rays are highly diagnostic of the species (Cione and
Torno, 1984; MS).
Repository: División Paleontología Vertebrados, Museo de La
Plata, La Plata, Buenos Aires Province.
Reference: unpublished.

PALAEONTOLOGICAL RECORD: URUGUAY (UNNAMED LITHOSTRATIGRAPHIC
UNIT

Localities: Punta Palmar, Vuelta del Palmar, Balneario La
Esmeralda, all in the coast of Rocha Department, Uruguay.
The remains have been found on the surface.
Material: pharyngeal teeth. N°669, 754, 755, 775, 782.
Repository: Departamento de Paleontología, Museo Nacional de
Historia Natural, Montevideo, Uruguay.
Reference: Mones (1976), Perea and Ubilla (1981).
Associated marine fishes: **Eugomphodus** sp, **Carcharhinus** sp,
Myliobatis sp.

ZOOGEOGRAPHICAL AND ENVIRONMENTAL CONSIDERATIONS

The sediments containing the fossil remains herein described,
correspond to a marine transgression. This work is

76

adequately based on geological and biological evidence (Fidalgo, 1979, and those papers cited in it; Parodiz, 1962; Tonni and Cione, 1984). However, nowadays, most of the considered area is occupied by the great estuary named Río de la Plata, related to the Paraná-Uruguay river systems.

The extension of the marine and/or brackish conditions in the Holocene of the area is tested by analizing the fish evidence. The ancient distribution of **P. cromis** corroborates and provides a more accurate delimitation of the marine and/or brackish conditions.

Some physical data about the present conditions of the Río de la Plata estuary, the geographical distribution of **P. cromis** and comments on its ecology are given below, in order to be compared with the fossil record, as well as some other fish evidence.

Urien (1972) (see Figure 2) provides a zonation of the Río de la Plata according to its salinity:
1. Inner-fluvial, frontal delta platform on the upper part of the estuary with fresh-water predominance;
2. Intermediate fluvial, extending from the imaginary line that links La Plata (Argentina) with Colonia (Uruguay) in the western part, to that one between Punta Piedras (Argentina) and Montevideo (Uruguay) towards the east. Though predominantly fluvial, brackish waters may appear towards the eastern limit;
3. Fluvio-marine, including the outer part of the estuary up to its eastern border, with predominant marine conditions but some fluvial influence;
4. Marine, normal marine conditions.

Souto (1974) does not differentiate a fluvio-marine from a marine zone, based on tintinid distribution. Otherwise, Boltovskoy and Lena (1978) confirm such a zonation based on foraminifers. In this zone, if the easterly and south-easterly winds are strong and constant, the brackish waters penetrate mainly along the bottom up to the western border of the fluvio-marine zone. Undoubtedly, the western part of the intermediate fluvial zone has a lower salinity than the eastern part of it, in which oscillations between 0.5 and 25 per thousand have been recorded (Boltovskoy and Lena, 1978).

The salinity is notably variable in the intermediate fluvial and fluvio-marine zones, as it is shown by the following data:
1. Montevideo Bay, 1947, from 1.63 per thousand to 33-69 per thousand (De Buen, 1950, **fide** Boltovskoy and Lena, 1978). In 1928, values ranging from 7.40 to 24.42 g per thousand were

77

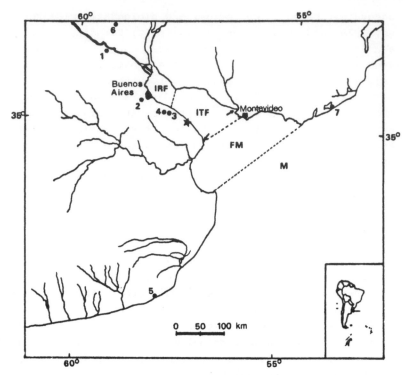

Figure 2. Localities (see text) and Río de la Plata zonations
by Urien (1972). IRF: Inner-fluvial; ITF: Intermediate-fluvial;
FM: Fluvio-marine; M: Marine. Arrows indicate Punta Piedras,
Argentina and mouth of Río Santa Lucía, Uruguay. Star indicates
Magdalena, Argentina.

registered (Ringuelet **et al.**, 1967). Between 1937 and 1947,
the fluctuations are significant though smaller than 27.19 g
per thousand (Sierra, 1974).
2. Pontón Recalada, from 0 to 35 g per thousand for surface
values (Ottmann and Urien, 1965).
3. Samborombón Bay, with maxima of 16 g per thousand at the
bottom and 7 g per thousand at the surface (Urien, 1967).
Menni (1984) obtained summer salinity values from 5.23 g per
thousand to 8.41 g per thousand. Urien (1972) reported higher
values of up to 25 g per thousand near the bay.
   Sciaenidae include marine and some continental species, in
warm-temperate or warm waters. Two groups are distinguishable
in the Argentinian species: one of them living near the coast
and another in rather offshore, but not in very deep waters.

In the first group, **Micropogonias furnieri** and **P. cromis** can be found, whereas in the second, **Macrodon ancylodon** and **Cynoscion striatus** have been cited. The former two penetrate in estuaries. In the Argentinian coast, juveniles of **M. furnieri** reach the city of Buenos Aires (López and Castello, 1967); on the other hand, **P. cromis** does not enter so much in the estuary. Adults reach the proximity of Punta Piedras, north of Samborombón Bay, according to commercial fishermen of La Boca, **fide** J. Mestre and L. Braga (pers. com., see Figure 2). This last species has its southern limit in Golfo San Matías (lat. 41°-42° S). In the Uruguayan coast, big black drums are recorded up to the mouth of Río Santa Lucía, westwards from Montevideo (**fide** C. Ríos and R. Vaz Ferreira, pers. com.). **P. cromis** is tolerant to a wide range of temperature and salinity (Simmons and Breuer, 1962). It is omnivorous and reaches a maximum reported size of 66.28 kg (Richards, 1973). This author mentions captures in the range of 9 °C in U.S. waters.

P. cromis is characteristic of the Argentinian zoo-geographical Province. In Punta Piedras this species is recorded with other warm-temperate or warm marine, amphibiotic and some continental species, such as **Mugil lisa, Lycengraulis olidus, Macrodon ancylodon, Luciopimelodus pati, Parapimelodus valenciennesi**, among others. **P. cromis** is usually present from September to March (**fide** L. Braga).

Braga (1984) studied its feeding habits, which are related almost exclusively to the bentonic complex. Preadults and adults are malacophagous. Specimens captured by commercial fishermen of San Clemente del Tuyú, south of Bahía Samborombón show in their stomachs remains of **Mactra marplatensis** (Bivalvia) and **Littoridina piscium** (Gastropoda), while those specimens obtained at Punta Piedras coast show in their stomachs a content of **Erodona mactroides** (Bivalvia) and **Littoridina piscium**. Consequently, it is evident that this species has not specific taxonomic preferences. Moreover, **E. mactroides** reaches further northern localities, v. gr. Magdalena (**fide** E. Martín, pers. com.) where **P. cromis** is unknown. These facts confirm that the range of this fish is not limited in the estuary by feeding habits (Figure 2).

Turbidity is higher in Punta Piedras than in more northern localities (**fide** L. Braga), suggesting that this factor is here not relevant to the distribution of **P. cromis** (Figure 2).

It is therefore apparent that turbidity and feeding factors

would not have any decisive influence on the distribution of
P. cromis. This species reaches areas of variable salinity;
however, it is not recorded neither in the fluvial zone, nor
in most of the intermediate fluvial zone. Consequently, a low
salinity environment is determinant of its range.

The hypothesis proposed in this paper is that salinity in
localities 1, 2, 3, 4 and 6 was during Mid-Holocene at least
similar to that presently recorded at Punta Piedras and the
mouth of Río Santa Lucía or even more marine. This implies
that salinity conditions have corresponded to those recorded
today at the limit between intermediate fluvial and fluvio-
marine zones or were even higher. This is corroborated by the
present distribution of **P. cromis**, which is interpreted as
a consequence of salinity as the main limiting factor, and
the record of other marine fishes (see Cione, 1978; Tonni and
Cione, 1984). Localities 5 and 7 are presently at the marine
coast and similar conditions of salinity are inferred for
earlier Holocene times at these sites.

ACKNOWLEDGEMENTS

We are grateful to Drs. M. M. Azpelicueta, F. Fidalgo and
E. P. Tonni for the critical reading of the manuscript. We
are also indebted to Drs. L. Braga, J. Mestre, C. Ríos and
R. Vaz Ferreyra for important data concerning black drum
distribution, Dr. A. Mones for loaning the Uruguayan material,
Mr. A. Aliscioni and Dr. M. Bond for the Ezeiza specimen,
Dr. J. Bonaparte, Mr. J. Petrocelli and Dr. E. Rossi for the
authorization to examine their collections, and Dr. E. Martín
for data on invertebrate zoogeography.

REFERENCES

Ameghino, F. 1898. Sinopsis geológica-paleontológica. **In:**
    Segundo Censo de la República Argentina, 1:111-255. Buenos
    Aires.
Boltovskoy, E. and Lena, H. 1978. Foraminíferos del Río de la
    Plata. **Boletín, Servicio de Hidrogafía Naval**. 661:1-22.
    Buenos Aires.
Braga, L. 1984. Contribución al conocimiento de la alimentación
    de los Sciaenidae y las adaptaciones del aparato digestivo

a sus respectivos regímenes. Tesis doctoral, Fac. Cienc. Nat. Museo, Universidad Nacional de La Plata, unpublished. La Plata.

Cione, A.L. 1978. Aportes paleoictiológicos al conocimiento de las paleotemperaturas en el área austral de América del Sur durante el Cenozoico. Aspectos zoogeográficos y ecológicos conexos. **Ameghiniana**, 15(1-2):183-208. Buenos Aires.

Cione, A.L. and Torno, A.E. 1984. Descripción y comparación de peculiares estructuras en la aleta dorsal de **Pogonias cromis** (Perciformes, Sciaenidae) y de una especie de la misma familia del Terciario marino de Entre Ríos, Argentina. Resumenes de la VII Jornadas Argentinas de Zoología: 178. Mar del Plata.

Cione, A.L. and Torno, A.E. **MS**. Dorsal fin structures of some Recent and fossil Sciaenids (Perciformes) of Argentina. Ontogenetic modifications and phylogenetic and adaptive significance.

Fidalgo, F. 1979. Upper Pleistocene-Recent marine deposits in Northeastern Buenos Aires province (Argentina). International Symposium on coastal evolution in the Quaternary, São Paulo, p.384-404.

López, R. and Castello, J.P. 1967. Corvinas del Río de la Plata. **Boletín del Servicio de Hidrografía Naval**, 5(1):14-27. Buenos Aires.

Menni, R. 1983. Los peces en el medio marino. Editorial Sigma, Buenos Aires, 169pp.

Mones, A. 1976. Notas paleontológicas uruguayas, III. Vertebrados fósiles nuevos o poco conocidos (Chondrichthyes, Osteichthyes, Amphibia, Mammalia). **Ameghiniana**, 12(4):343-349. Buenos Aires.

Ottmann, F. and Urien, C. 1965. Le mélange des eaux douces et marines dans le Río de la Plata. **Cahiers oceanographiques**, 17(10):703-713.

Parodiz, J.J. 1962. Los moluscos marinos del Pleistoceno rioplatense. **Comunicaciones Sociedad Malacológica de Uruguay**, 1(2):29-46.

Perea, D. and Ubilla, M. 1981. Estudio preliminar de la ictiofauna fósil marina de las costas del Departamento de Rocha, Uruguay. Resúmenes de Comunicaciones de las Jornadas de Ciencias Naturales, 2:25-26. Montevideo.

Richards, C.E. 1973. Age, growth and distribution of the black drum (**Pogonias cromis**) in Virginia. **Transactions of the American Fisheries Society**, 3:584-590. Washington.

Ringuelet, R.; Arámburu, R. and Alonso de Arámburu, A. 1967.
Los peces argentinos de agua dulce. CIC, La Plata.

Sierra, B. 1974. Caracteres métricos de **Blennius fissicornis**
Quoy & Gaymard, 1824, correlacionados con la salinidad del
Río de la Plata (Teleostei, Blennidae). **Physis**, A33(86):
347-350. Buenos Aires.

Simmons, E.G. and Breuer, J.P. 1962. A study of red fish
Sciaenops ocellata L. and black drum, **Pogonias cromis** L.
**Publications of the Institute of Marine Science, University
of Texas,** 8:184-211.

Souto, S. 1974. Tintínidos del Río de la Plata y su zona de
influencia (Protozoa, Ciliata). **Physis**, A33(87):201-205.
Buenos Aires.

Tonni, E. and Cione, A.L. 1984. A thanatocenosis of continental
and marine vertebrates in the Las Escobas Fm (Holocene) of
northeastern Buenos Aires province, Argentina. **Quaternary
of South America and Antarctic Peninsula,** 2:93-113.
Rotterdam.

Tonni, E. and Fidalgo, F. 1978. Consideraciones sobre los
cambios climáticos durante el Pleistoceno tardío-Reciente
en la provincia de Buenos Aires. Aspectos ecológicos y zoogeográficos relacionados. **Ameghiniana,** 15(1-2):235-253.
Buenos Aires.

Urien, C. 1967. Los sedimentos modernos del Río de la Plata
exterior, Argentina. **Boletín del Servicio de Hidrografía
Naval,** 4(2):113-213. Buenos Aires.

Urien, C. 1972. Río de la Plata estuary environments. **In:**
Nelson S. (ed.), Environmental framework of coastal plain
stuaries. **Geological Society of America, Memoir,** N°133.
Washington.

HÉCTOR O.PANARELLO

Instituto de Geocronología y Geología Isotópica (INGEIS), Buenos Aires, Argentina

6

Oxygen-18 temperatures on present and fossil invertebrated shells from Túnel Site, Beagle Channel, Argentina

ABSTRACT

First isotopic temperatures obtained at INGEIS on present and fossil shells of marine invertebrates at Túnel Site, Beagle Channel, are presented.

$^{18}O/^{16}O$ ratios were measured on carbonatic shells of the species **Mytilus edulis**, **Nacella deaurata**, **Fissurella maxima**, **Tawera gahi**, **Aulacomya ater ater**, **Trophon sp** and **Balanus sp.**

The analyzed samples were collected from living specimens and from two well-dated middens, whose carbon-14 ages are:

$\propto$ level: 3030 ± 100 yr BP

7D level: 5630 ± 120 yr BP

In most cases, samples of the seven mentioned species were available for the three moments considered.

Results show that the sea-water temperature ca 5600 yr BP (7D level) was lower than the present one, probably reflecting local phenomena such as marine stream pathway changes and not a world-wide trend, characterized by higher temperatures at that time. Conversely, isotopic data at 3030 yr BP ($\propto$ level) suggest higher temperatures than the present ones.

RESUMEN

Como parte de un estudio paleoclimático más amplio, se presentan en este trabajo las primeras temperaturas isotópicas obtenidas en el INGEIS sobre muestras fósiles y actuales en el sitio Túnel, Canal de Beagle.

Se determinó la relación $^{18}O/^{16}O$ en el caparazón carbonático de las especies **Mytilus edulis**, **Nacella deaurata**, **Fissurella**

maxima, **Tawera gahi, Aulacomya ater ater, Trophon** sp y
Balanus sp.

Las muestras analizadas corresponden a organismos vivientes
en la zona y a ejemplares hallados en dos sitios arqueológicos
cuyas edades carbono-14 son las siguientes:

nivel ∝ : 3030 + 100 a A.P.

nivel 7D: 5630 + 120 a A.P.

En la mayoría de los casos, las especies citadas se encontra-
ban en conjunto en cada uno de los niveles estudiados.

De acuerdo a los resultados, la temperatura del agua del mar
correspondiente al nivel 7D, 5630 a A.P., habría sido más baja
que la actual, en tanto la del nivel ∝ , habría sido más alta.

Los valores térmicos del nivel ∝ están de acuerdo con los
lineamientos climáticos globales. Por su parte los del nivel
7D, muestran una situación de enfriamiento en contradicción
con la tendencia mundial. Este período de bajas temperaturas,
podría interpretarse como producido por fenómenos locales hacia
5630 a A.P., como podría ser el acercamiento de una corriente
marina fría a la zona del canal.

INTRODUCTION

After the development of the carbonate-water temperature scale
by Epstein **et al.** (1951, 1953), based on the oxygen isotope
fractioning, many reports have been published all over the
world applying this tool to palaeoclimatic, glaciological and
oceanographic studies.

The recent development of stable isotope ratios measurements
in Argentina (Panarello **et al.**, 1981; Linares **et al.**, 1982;
Panarello and Parica, 1984) leads to the determination of the
evolution of palaeotemperatures over marine shells collected
at the Beagle Channel.

The temperature differences between two well-dated
settlement stages of an archaeological site and the present
values are intended to be established in this work. For this
purpose, 20 samples comprising the two levels and living
specimens were collected in order to compare the temperatures
of the 3 moments for 7 different species.

^{13}C and x-ray diffraction measurements have also been
performed to ensure crystalline structure preservation.

Present isotopic temperatures are compared with those
recorded with a thermograph located at a depth of 1.30 m in
the Beagle Channel, at Ushuaia, near Túnel Site.

WORKING AREA, MATERIALS, METHODS AND TECHNIQUES

Túnel Site is located on the north shore of the Beagle Channel, Isla Grande de Tierra del Fuego, at lat. 54°49'15" S and long. 68°09'44" W. The climate is cold and rainy.

Samples were obtained from levels α and 7D corresponding to the archaeologic excavation performed by Orquera and Piana (1983). The original nomenclature for these levels is here presented.

α level is a shell midden with interbedded charcoal. The carbon-14 age of this charcoal, which has been assumed as "true age" for the midden, is 3030 ± 100 yr BP (Piana, 1984).

The 7D level bears several different datable archaeological remains proved to be coetaneous. Carbon-14 ages on these materials are listed in Table 1, (Albero et al., 1986).

Table 1. Radiocarbon ages on several materials of the 7d level (after Albero et al., 1986)

MATERIAL	SPECIES	^{13}C (% PDB)	^{14}C age (yr BP)
Charcoal	Nothofagus sp	-24.0	5630 ± 120
Shell	Nacella sp	3.4	6220 ± 120
Shell	Fissurella sp	2.8	6340 ± 120
Shell	Balanus sp	3.0	5110 ± 130
Shell	Mytilus sp	2.1	6180 ± 120
Shell	Aulacomya sp	---	6290 ± 120
Shell	Trophon sp	4.9	5720 ± 120
Collagen (bone)	Lama guanicoe	-20.5	5280 ± 100
Collagen (bone)	Arctocephalus australis	-11.8	6220 ± 140

The assumed "true age" for this level, according to these authors, is that of the charcoal: 5630 ± 120 yr BP.

Samples of Mytilus edulis, Nacella deaurata, Fissurella maxima, Aulacomya ater ater, Tawera gahi, Balanus sp and Trophon sp, have been collected from the α , 7D levels and for channel living specimens, lacking only one sample of T. gahi.

All the samples were collected and classified by Lic. E. Piana from the Centro Austral de Investigaciones Científicas (CADIC).

Table 2. Isotopic, X-ray mineralogical composition and
calculated temperatures (assumed $\delta w = -0.85$).

SPECIE	LEVEL	CALCITE+ ARAGONITE	$\delta^{13}C*$	$\delta^{18}O*$	T**
Mytilus edulis	present	5/95	0.8	0.3	11.6
	$\propto$	5/95	1.3	-0.1	13.2
	7D	15/85	1.1	-0.6	10.5
Nacella deaurata	present	100/0	2.4	1.0	9.0
	$\propto$	100/0	1.9	0.9	9.4
	7D	100/0	1.9	1.2	8.0
Fissurella maxima	present	100/0	1.9	0.9	9.4
	$\propto$	95/5	1.2	0.9	9.4
	7D	55/45	2.0	1.2	8.3
Balanus sp	present	100/0	0.4	1.4	7.3
	$\propto$	100/0	1.7	1.0	9.0
	7D	100/0	1.9	1.8	6.2
Aulacomya ater ater	present	14/86	2.1	1.0	9.0
	$\propto$	---	0.7	0.25	13.9
	7D	0/100	2.4	1.0	9.0
Trophon sp	present	100/0	2.8	1.7	6.4
	$\propto$	100/0	2.8	1.4	7.6
	7D	100/0	3.5	1.9	5.9
Tawera gahi	present	10/90	0.8	-0.15	13.5
	$\propto$	not available			
	7D	0/100	0.8	0.3	11.6

```
 *    ‰  ± 0.1 ‰   vs PDB
**    °C ± 0.5 °C
 +    %  ± 15 %
```

Shells were first scrapped, cleaned and washed with 10% HCl,
then crushed and grinded to 80-100 mesh and finally heated at
400°C **in vacuo** (10^{-2} Pa) during about 30 minutes in order
to destroy organic matter.

After this treatment, samples were reacted with 100% H_3PO_4
as described in Mc Crea (1950).

The $^{13}C/^{12}C$ and $^{18}O/^{16}O$ ratios were measured against Carrara

Marble CO_2 reference in a Micromass 602-D double collector, McKinney type mass spectrometer, as described in Panarello et al. (1980).

Results are expressed in δ (‰) defined as:

$$\delta = \frac{Rs}{Rstd} - 1 \quad x \ 1000\%$$

where:

$\delta = \delta^{13}C$ or $\delta^{18}O$: isotopic deviation in per thousand
$R = {}^{13}C/{}^{12}C; \ {}^{18}O/{}^{16}O$: isotope ratios
s = sample
std= standard

Isotopic temperatures were computed taking into account the scale proposed by Epstein et al. (1953):

$$t = 16.4 - 4.2 \ (\delta c - \delta w) + 0.13 \ (\delta c - \delta w)^2 \ (°C)$$

being δ c and δ w oxygen isotope composition of the shell carbonate and the water from where it precipitated, respectively.

Water $^{18}O/^{16}O$ ratios (δ c) were determined accordingly to Epstein and Mayeda (1953) and Roether (1970), being 0.85 ± 0.05 ‰ and assumed constant for all samples. This water was collected near the termograph, in the Beagle Channel.

RESULTS

Table 2 shows the $\delta^{13}C$ and $\delta^{18}O$ values for the three moments for the seven species. In addition, the mineralogical composition, previously determined by X-ray diffraction, is presented.

As it can be seen, despite the isotopic temperatures are not consistent, a systematic trend is evidenced in most of the cases:

T (7D) < T (present) < T (α)

The "t" Student test, $t = (d - 0) /s \ n^{1/2}$ in the zero point where:

d: mean value of the difference T (α) - T (present)
s: 0.75; standard deviation estimate
n: 5; number of pairs of samples (with the exception of **A. ater ater** since it is rather erratic) gives t = 2.92, outside the critical interval $t_{0.05;4}$ = 2.57 (2 tails).

Applied to the T (present) - T (7D) difference with d = 1.12, s = 0.45 and n = 6 (with the exception of **A. ater ater**) gives t = 6.1, outside the critical interval $t_{0.002;5}$ - 5.89 (2 tails). Therefore, the null hypothesis is rejected in both cases and temperatures are significantly different:

T ($\propto$) - T (present) = 0.98 $\pm$ 0.75 °C

T (7D) - T (present) =-1.12 $\pm$ 0.45 °C

Values obtained on shells belonging the $\propto$ level (3100 yr BP) agree with the global trend of higher temperatures for this period (Mörner, 1984).

As it was pointed out before, 7D shells (5630 yr BP) show isotopic temperatures lower than present ones, in disagreement with Northern Hemisphere data. Moreover, in a recent work carried out by González **et al.**(1984) on isotope ratios on **Tagelus gibbus** shells collected at Bahía Blanca estuary ($\cong$ 38° S; 67° W), the "Hypsithermal" was found at 6000 radiocarbon yr BP. Taking into account i) that this age is not reservoir-effect corrected, and ii) its analytical error, this moment could be considered isochronous with the formation of the 7D level. The Hypsithermal was characterized by climate improvement and maximum fresh water input to this estuary, as it was indicated by the isotope depletion of both oxygen-18 and carbon-13. These facts suggest that the lowering of the Beagle temperatures at 5630 radiocarbon yr BP was due to local circumstances.

Although carbon-13 values do not show a systematic trend, five over seven species show major isotope enrichment on 7D samples. The average values appear as follows:

	PRESENT	$\propto$	7D
$\delta^{13}C$ (‰) Average	1.9	1.9	2.3

In other words:

$$\delta^{13}C_{present} \delta = {}^{13}C_{\propto} < \delta^{13}C_{7D}$$

Therefore, the continental-marine waters ratio in 3030 ^{14}C yr BP was similar to the present one and that of 5630^{14}C yr BP was lower. If this is true, $\delta^{18}O_w$ at 5630 yr BP would be greater than the assumed $\delta^{18}O_w$ = -0.85 ‰ and consequently, temperatures as those calculated by the Epstein **et al.**'s (1951, 1953) equation increased. This fact would vanish the difference with present data.

The remaining question is: why if 7D temperatures were actually higher, fresh-water coming from glacier melt had decreased?

Because of the 7D isotope composition: high $\delta^{18}O$ and $\delta^{13}C$, it could be inferred that its temperature was actually lower, but probably due to a local process such as the approaching of a cold marine stream to the Channel. Further studies are necessary either to confirm or to change this hypothesis.

Concerning individual values of temperatures, it may be regarded that:

Present temperatures of 7.3 °C and 6.4 °C measured on **Balanus sp** and **Trophon sp,** respectively, agree with the average value obtained in the Thermograph (6.0 ± 1.7) °C. Thus, these species seem to be really useful for recording absolute thermal values in the Channel.

N. deaurata, F. maxima and **A. ater ater** yield oxygen-18 temperatures rounding 9 °C, probably because they are recording only the warmer periods along the year. This may be a consequence of an increasing growth-rate of these species for temperatures over 6 °C.

Higher temperatures showed by **T. gahi** cannot be explained in the same way. This clam lives buried in sediments where chemical reactions of degrading organic matter increase the local temperature, specially in the marine littoral at high latitudes, where the coastal biomass is rather large.

α - **A. ater ater** value is abnormally high. This fact, together with its low $\delta^{13}C$ value, leads to suspect the alteration or contamination of the available shell samples.

CONCLUSIONS

Oxygen-isotope temperatures are presented as a reliable tool in paleotemperature studies in Túnel Site, Beagle Channel.

Significative differences have been observed among present temperatures and those corresponding to α and 7D levels.

α - mean temperatures would probably have been 1 °C above the present ones, whereas 7D temperatures would have been lower.

α - mean temperatures agree with the world-wide trend of warm climate at about 3000 yr BP, while 7D ones seem to be caused by local processes, such as the approaching of a cold marine stream.

Fixed, buried living clams like **Tawera gahi** are not

recommended for temperature studies.

Trophon sp and **Balanus sp** are the species that record the real temperature more accurately in the channel waters.

ACKNOWLEDGEMENTS

The author is indebted to E. Piana who collected and classified the samples for this study and to the staff members of the Stable Isotope Laboratory of the INGEIS. He is also very grateful to R. Gonfiantini, M.C. Albero and E. Linares for their valuable suggestions.

This work is the Contribution N° 98 of the Instituto de Geocronología y Geología Isotópica (INGEIS).

REFERENCES

Albero, M.C., Angiolini, F.E. and Piana, E.L. 1986. Discordant ages related to reservoir effect of associated archeological remains from Túnel Site, Beagle Channel, Argentina. **Radiocarbon**, 28(24):748-753.

Epstein, S., Buchsbaum, R., Lowenstam, H. and Urey, H. 1951. Carbonate-water isotopic temperature scale. **Geol. Soc. America. Bull.** 62:417-426.

Epstein, S., Buchsbaum, R., Lowerstam, H. and Urey, H. 1953. Revised carbonate-water isotopic temperature scale. **Geol. Soc. America. Bull.** 64:1316-1326.

Epstein, S. and Mayeda, T.K. 1953. Variation of $\delta^{18}O$ content of water from natural sources. **Geochim. Cosmochim. Acta.** 4:213-224.

González, M.A., Panarello, H.O., Marino, H. and Valencio, S.A. 1984. Niveles marinos del Holoceno en el estuario de Bahía Blanca, Argentina. Isótopos estables y microfósiles como indicadores paleoambientales. **In.** E. Schnack (ed.), Simposio "Oscilaciones del nivel del mar durante el último hemiciclo glacial en Argentina. p.48-68. Mar del Plata, Argentina.

Linares, E., Panarello, H.O., Valencio, S.A. and Garcia, C.M. 1982. Isótopos del carbono y oxígeno y el origen de las calizas de las Sierras Chica de Zonda y de Pie de Palo, Provincia de San Juan. **Asoc. Geol. Argentina Rev.** 37(1):80-90. Buenos Aires.

McCrea, J.M. 1950. The isotopic chemistry of carbonates and a paleo-temperature scale. **Journ. Chem. Phys.** 18:849-857.

Mörner, N.A. 1984. Planetary, solar, atmospheric, hydrospheric and endogene processes as origin of climatic changes on the earth. **In:** Climatic Changes on yearly to Millennial Basis. N.A. Mörner and W. Kárlen (eds.) D. Reidel Publishing Company. Dordrech, Boston, Lancaster: 483-507.

Orquera, L.A. and Piana, E.L. 1983. Prehistoric maritime adaptation at the Magellan Fueguian littoral. Submitted at New World Maritime Adaptation Symposium, 48th. Annual Meeting of the Society for American Archaeology, Pittsburg. April 30, 1983.

Panarello, H.O., Garcia, C.M., Valencio, S.A. and Linares, E. 1981. Determinación de la composición isotópica del carbono en carbonatos. Su aplicación a la Hidrogeología y Geología. **Asoc. Geol. Argentina Rev.**, 35:460-466. Buenos Aires.

Panarello, H.O. and Parica, C.A. 1984. Determinación de la composición isotópica del oxígeno. Primeros valores en aguas de lluvia de Buenos Aires. **Asoc. Geol. Argentina Rev.**, 39(1-2):3-11. Buenos Aires.

Piana, E.L. 1984. Arrinconamiento o adaptación en Tierra del Fuego. **In:** Universidad de Belgrano (ed.), Ensayos de antropología argentina. p.5-110. Buenos Aires.

Roether, W. 1970. Water CO_2 exchange set up for the routine oxigen-18 assay of natural water. **Int. Journ. App. Rad. and isotopes**, 21:379-387.

CALVIN J.HEUSSER
New York University, USA

7

Fire history of Fuego-Patagonia

ABSTRACT

Incidence of late Quaternary fires in Fuego-Patagonia is
interpreted from amounts of charcoal (particles 50-150 um x
cm^{-3}) found in five geographically distinct sedimentary
sections. The southernmost section (940 cm) is of a mire on
Isla Navarino (54°56'S, 67°38'W) in deciduous beech (**Nothofagus
pumilio**) forest. Two sections (760 and 800 cm) are of mires
bordering the Strait of Magellan at Puerto del Hambre (53°36'S,
70°55'W) in mixed deciduous beech-evergreen beech (**N.
betuloides**) forest and at Punta Arenas (53°09'S, 70°57'W) in
an ecotone of steppe (**Festuca gracillima - Chiliotrichum
diffusum**) - deciduous beech forest. The northernmost sections
(850 and 29 cm) at Torres del Paine are of a summer-dry lake
in steppe (50°59'S, 72°40'W) and a mire (51°05'S, 73°04'W) in
the forest-steppe ecotone. Sections range in age between
15,800 and 660 yr BP with their chronology controlled by a
total of 27 C-14 determinations.

Fires during the late glacial were apparently minor and
infrequent, except at Torres del Paine. Their intensity and
frequency on the landscape were evidently greatest in the
early Holocene (23,000-32,400 charcoal particles cm^{-3}) when
pollen records show the extent of steppe-tundra reduced by the
spread of beech communities. On Isla Navarino, fires were least
frequent and comparatively insignificant (average <1000 charcoal
particles cm^{-3}) compared with other section sites. A detailed
charcoal-pollen record of the short section at Torres del Paine
covering recent centuries shows the impact of fire on
vegetation in the inverse relationship between amounts of
charcoal and beech. Following initial European settlement and

the establishment of Parque Nacional Torres del Paine, fire
influence steadily declined accompanied by an increase of beech.

Fire are attributed to paleo-Indian hunters, European
settlers, and volcanic activity. Charcoal associated with human
artifacts in cave deposits as old as 12,600 yr BP in the
Patagonian steppe suggests that hunting parties were also the
cause of late glacial burning recorded at Torres del Paine on
the western boundary of the steppe. Times of vulcanism, as
revealed by major, widespread layers of tephra, occurred at
9400-8900, 6600-4500, and 2200 yr BP, and peak amounts of
charcoal between 10,000-9000 and around 6000 yr BP in the
sections are in time-stratigraphic agreement with ages of two
of the regional tephra layers. Volcanic centers are at Mt.
Burney in Pacific coastal Chile (52°20'S) and the field of
craters north of the Atlantic end of the Strait of Magellan in
Chile and adjacent Argentina. Lightning, virtually non-existent
in the region of the sections, is discounted as a cause of fire.

RESUMEN

La incidencia de incendios del Cuaternario Superior en Tierra
del Fuego y Patagonia se interpreta a partir de partículas de
carbón (de 50 a 150 um x cm^{-3}) halladas en cinco secciones
sedimentarias geográficamente distintas. La sección más meridional (940 cm) es de un turbal en la Isla Navarino (54°56'S,
67°38'W) en bosque de fagáceas caducifoliadas (**Nothofagus
pumilio**). Dos secciones (760 y 800 cm) proceden de turbales
que bordean el Estrecho de Magallanes en Puerto del Hambre
(53°36'S, 70°55'W) en bosque mixto de fagáceas caducifoliadas
y perennifoliadas (**N. betuloides**) y en Punta Arenas (53°09'S,
70°57'W) en un ecotono de estepa (**Festuca gracillima -
Chiliotrichum diffusum**) - bosque de fagáceas caducifoliadas.
Las secciones más septentrionales (850 y 29 cm) en Torres
del Paine son de un lago (seco en el verano) en la estepa
(51°05'S, 73°04'W) en el ecotono bosque-estepa. Las secciones
varían en edad entre 15.800 y 660 años antes del presente,
con su cronología controlada por un total de 27 determinaciones
radiocarbónicas.

Los incendios durante el final de la época glacial fueron
aparentemente menores e infrecuentes, excepto en Torres del
Paine. Su intensidad y frecuencia sobre el paisaje fueron
evidentemente mayores en el Holoceno temprano (23.000 a 32.400

partículas de carbón por cm^{-3}) cuando los registros polénicos
muestran que la extensión de la estepa-tundra había sido re-
ducida por la expansión de las comunidades de fagáceas. En
Isla Navarino, los incendios fueron menos frecuentes y comparativamente insignificantes (promedio <1000 partículas de carbón
por cm^{-3}) comparado con otras secciones. Un registro polínico
detallado de una sección corta en Torres del Paine cubriendo
los siglos más recientes muestra el impacto del fuego en la
vegetación, en forma de relaciones inversas entre cantidades
de carbón y de fagáceas. Luego del establecimiento inicial de
los europeos y la creación del Parque Nacional Torres del Paine,
la influencia de los incendios declinó fuertemente, acompañada
por un aumento de las fagáceas.

Los incendios son atribuídos a cazadores paleo-indios, colo-
nos europeos y actividad volcánica. La existencia de carbón
asociado con artefactos humanos en depósitos de cuevas tan
antiguas como 12.600 años AP en la estepa patagónica sugiere
que grupos de caza habrían sido también la causa de quemazones
del Glacial Tardío, registradas en Torres del Paine sobre el
límite occidental de la estepa. Epocas de vulcanismo, como ha
sido revelado por capas de tefra importantes y muy extendidas,
tuvieron luego en 9400-8900, 6600-4500 y 2200 años AP, y
cantidades máximas de carbón entre 10.000-9000 y alrededor de
6000 años AP, hallados en las secciones, están en concordancia
cronoestratigráfica con edades de dos de las capas téfricas
regionales. Los centros volcánicos son el Monte Burney en la
costa pacífica chilena (52°20'S) y el campo de cráteres al
norte del extremo atlántico del Estrecho de Magallanes en Chile
y la Argentina adyacente.

Los relámpagos, virtualmente inexistentes en la región donde
están ubicadas las secciones, son descartados como causa de
incendio.

INTRODUCTION

When Hernando de Magallanes traveled the coast of southernmost
South America during his circumnavigation of the goble in
1520 A.D., he named the archipelago south of the strait through
which he sailed, Tierra del Fuego, or "land of fire", from the
many fires burning on shore; three centuries later in 1833 A.D.,
"fires...lighted on every point" in the southern part of Tierra
del Fuego likewise attracted the attention of Charles Darwin

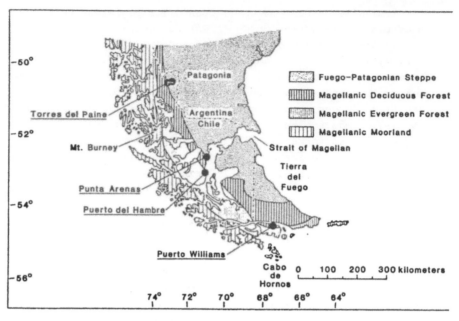

Figure 1. Location in Fuego-Patagonia of section sites
(underlined) and physical features referred to in the text.
Major vegetation zones (Pisano, 1981; Moore, 1983) are shown
generalized (scattered areas of Andean tundra and barren
ground omitted)·

during the famed voyage of H. M. S. Beagle (Darwin 1958;
Godley, 1965). Fires set by Indians served as a means of
communication, but perhaps most important, as a source of
warmth in an inhospitable land of cold and powerful wind
during the time of the Little Ice Age in the Southern
Hemisphere (Lamb, 1977).

European settlement after about 1870 A.D. (Martinić, 1973,
1974, 1976, 1977, 1978) created fires in Tierra del Fuego
(Bridges, 1948) and, north of the Strait of Magellan, in
Patagonia (Grosse, 1974). Fires ignited in areas of forest,
lying in western and southwestern sectors of the region,
readily spread across the dry, eastern steppe, owing to the
strong, unceasing westerly wind. Before the discovery of
Fuego-Patagonia by European man, paleo-Indians used fire
(Bird, 1938). Charcoal found together with remains of lithic
industry in Patagonian cave deposits is as old as 12,600 yr BP
(Cardich et al., 1973; Cardich, 1978).

The purpose of this paper is to investigate the incidence
and regional distribution of past fires in Fuego-Patagonia
from the occurrence of charcoal particles in geographically
distinct, late Quaternary sedimentary sequences. Measurements
made in various parts of the globe on charcoal in lake
sediments (Davis, 1967; Swain, 1973, 1978; Singh **et al.**, 1981;
Macphail, 1984) and in soils (Payette and Gagnon, 1985;
Sanford **et al.**, 1985) indicate that fire has been a significant
and widespread environmental factor in the past. Fire data are
needed from Fuego-Patagonia to preclude erroneous
interpretation of Quaternary climate from vegetation changes
implied from fossil pollen records. Palynological work in the
region has given scant attention to fire, emphasizing climate
to account for vegetation change (Auer, 1958; Markgraf, 1983;
Heusser, 1984a).

LOCATION OF SEDIMENTARY SEQUENCES

Data are derived from five widely separated sections. The
southernmost section (Figure 1) near Puerto Williams on Isla
Navarino (54°56'S, 67°38'W) is of a mire (940 cm deep) located
in deciduous southern beech (**Nothofagus pumilio**) forest
(Pisano, 1977). Sections along the Strait of Magellan, also
mires (respectively, 760 and 800 cm deep), are at Puerto del
Hambre (53°36'S, 70°55'W) in mixed deciduous beech-evergreen
beech (**N. betuloides**) forest and at Punta Arenas (53°9'S,
70°57'W) in an ecotone of steppe (**Festuca gracillima-
Chiliotrichum diffusum**) - deciduous beech forest (Pisano, 1973,
1977) which before human clearance and settlement contained
greater coverage of deciduous beech (**N. antarctica**) forest or
woodland (Auer, 1958; Zamora, 1975). The first of the
northernmost sections (850 cm) at Torres del Paine (50°59'S,
72°40'W) is from a small summer-dry lake, situated near Lago
Sarmiento in steppe vegetation characterized by **Festuca
gracillima** (Pisano, 1974), a short distance beyond the east·
gate of Parque Nacional Torres del Paine. The second (29 cm)
is of a mire (51°05'S, 73°04'W) in the forest-steppe ecotone
to the west inside the national park·.
Sections of deposits in beech forest under existing climate
are located in a zone of 400-5500 mm average annual
precipitation (evergreen beech forest prevailing where there
is an excess of about 550 mm), whereas sections in forest/
steppe ecotone or steppe, are subject to <200-400 mm (Zamora

Table 1

Radiocarbon age determinations for sections from Fuego-
Patagonia

Section	Age and Laboratory number	
Puerto Williams	700 ± 60 (QL-1713)	10,080 ± 140 (QL-1718)
	2400 ± 40 (QL-1714)	10,510 ± 80 (QL-1684)
	3100 ± 60 (QL-1715)	11,850 ± 50 (QL-1720)
	3520 ± 60 (QL-1716)	12,730 ± 90 (QL-1685)
	5520 ± 70 (QL-1717)	
Puerto del Hambre	3970 ± 70 (QL-1467)	12,740 ± 260 (QL-1620)
	7980 ± 50 (QL-1468)	13,190 ± 80 (QL-1545)
	10,940 ± 70 (QL-1544)	15,800 ± 200 (QL-1469)
Punta Arenas	2330 ± 350 (QL-1546)	10,840 ± 70 (QL-1548)
	5940 ± 100 (QL-1547)	11,960 ± 170 (QL-1622)
	9240 ± 140 (QL-1621)	13,400 ± 140 (QL-1470)
Torres del Paine (long section)	3780 ± 150 (QL-1624)	7570 ± 200 (QL-1625)
	6380 ± 80 (QL-1550)	10,870 ± 70 (QL-1475)
	6870 ± 80 (QL-1553)	
Torres del Paine (short section)	660 ± 100 (QL-1549)	

---

and Santana, 1979; Prohaska, 1976; Miller, 1976). Overall,
precipitation shows no clear seasonal trends, although in
winter and spring, it is somewhat less than during the
remainder of the year. Wind in the latitudes of Fuego-
Patagonia ("screaming fifties") is both forceful and incessant,
averaging 10 m.sec^{-1} with maxima exceeding 30 m.sec^{-1}.

LABORATORY METHODS

Samples taken at 10-cm intervals from sections, except the
short section at Torres del Paine which was sampled at 1-cm
intervals, were processed in the laboratory (Heusser and
Stock, 1984). Quantities of charcoal (microscreened 50-150 um
in size) were counted under the microscope and measured
(particles.cm^{-3}) using marker pollen. Owing to the fact that
particle size and number are affected by differential chemical
treatment (Clark, 1984), a uniform processing schedule was
maintained for all samples. Plant cellular and structural
features, recognizable in scorched or blackened fragments,
distinguished charcoal from inorganic material. Charcoal
influx was not calculated because of uneven chronological
control from section to section (Table 1). For the short
section at Torres del Paine, counts of pollen and spores
were made along with charcoal to assess at close intervals
the relationship of fire to vegetation change during recent
centuries.

RESULTS AND DISCUSSION

1 CHARCOAL DIAGRAM

Several features are immediately apparent from the amount and
distribution of charcoal in the sections (Figure 2). Quantities
of charcoal in the late glacial are minimal, except at Torres
del Paine, and small amounts may represent redeposited
material derived from older Pleistocene deposits. In the early
Holocene, between approximately 9500 and 6000 yr BP, peak
values for the entire lengths of record are close to 10,000
particles cm^{-3} at Puerto del Hambre and between 23,000 and
32,400 particles cm^{-3} at Punta Arenas and Torres del Paine.
Late Holocene charcoal is most abundant in the last 2000 years

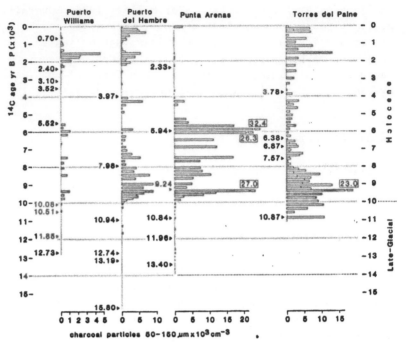

Figure 2. Distribution and abundance of charcoal particles in
late Quaternary sediments of sections from Fuego-Patagonia.
Data are adjusted to a scale of equal time, controlled by 26
radiocarbon age determinations (Table 1); note that scales
are not uniform. Boxed numbers indicate excessive values of
particles x 10^3 cm^{-3}.

with latest values covering the era of settlement by Europeans.
Evident in the Holocene are extended intervals of both charcoal
deposition and nondeposition, in some instances, as in the
case of Punta Arenas, reaching several millennia in length.
Torres del Paine shows what is evidently the lengthiest plot
of uninterrupted deposition. Quantities overall are lower in
plots of more humid forested sites (Puerto Williams and Puerto
del Hambre) than of sites in drier forest/woodland-steppe
(Punta Arenas and Torres del Paine).

Incidence and apparent intensity of past fires in Fuego-
Patagonia, as indicated by occurrence and amount of charcoal
in the representative sections, are attributed to paleo-
Indians, European settlers, and volcanic activity. Paleo-
Indian hunters conceivably are the principal cause of fires,
as burning seems certain to have been practiced to drive and

corral game. Cave deposits dated as early as 12,600 yr BP in the Patagonian steppe contain stone tools together with burnt bones of extinct horse, ground sloth, and camelid (Cardich et al., 1973; Cardich, 1978). This evidence for late glacial paleo-Indian use of fire may account for the singular quantity of 10,870 years old charcoal found at Torres del Paine at the edge of the steppe. A comparable occurrence is found in mid-latitude Chile for a charcoal peak dated at 11,380 yr BP associated with hunting of mastodon at Laguna de Tagua Tagua (Montané, 1968; Heusser, 1984); charcoal elsewhere in the region, as in Fuego-Patagonia, does not begin to appear in quantity until after 10,000 yr BP (Heusser, 1984b).

Volcanic eruptions constitute another possible cause of fire. Two centers of known vulcanism (Figure 1) are Mt. Burney (52°20'S, 73°25'W) in the west and the field of volcanic craters north of the eastern end of the Strait of Magellan in Chile and adjacent Argentina (Auer, 1959; Casertano, 1963; Skewes, 1978). Tephra layers, including those in the deposits of this study, owe their origin to three major eruptions dated at 9380 ± 90-8905 ± 110, 6600 ± 90-4480 ± 50, and 2240 ± 60 yr BP (Auer, 1958; Deevey et al., 1959).

Fires possibly caused by eruptions of Mt. Burney are conceivable sources for charcoal in any or all of the study records. Peak amounts of charcoal between 10,000-9000 and around 6000 yr BP in data plots (Figure 2) are in time-stratigraphic agreement with radiocarbon ages of two of the principal regional tephra layers. Less probable as a source than Mt. Burney is the volcanic area beyond the eastern end of the strait. Because of the westerly component of the wind, charcoal from fires in this area would not be expected to reach the section sites of this study. Only rarely, as meteorological data show (Zamora and Santana, 1979; Prohaska, 1976; Miller, 1976), does wind in Fuego-Patagonia blow from the east.

Lightning, a common means of ignition in many parts of the globe (Wright and Bailey, 1982), can be ruled out as a cause of fire in the region of Fuego-Patagonia where sections are located. Its occurrence is virtually unknown at high latitudes of South America in the belt of westerly winds (Prohaska, 1976; Miller, 1976).

Placed against a backdrop of pollen and spore stratigraphy (Heusser, 1984a), the plot of charcoal at Puerto Williams (Figure 2) shows little if any effect of fire on past vegetation. Under a warming climate in the early Holocene, for

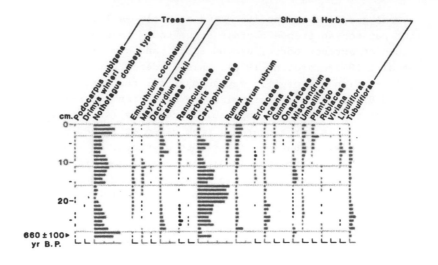

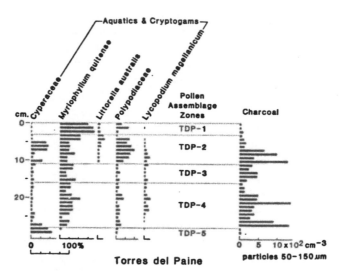

Torres del Paine

Figure 3. Pollen, spore, and charcoal diagram of the short section at Torres del Paine. Percentages of tree taxa and shrub and herb taxa are from pollen sums of at least 300 grains; percentages of aquatics and vascular cryptogams, calculated separately, are based on total counts of pollen (nonaquatic + aquatic) and spores.

example, there was an apparently rapid, unimpeded expansion of
beech in the tundra. At Puerto del Hambre, fluctuations of
charcoal and pollen taxa are more obvious, although somewhat
complicated by an episode of middle Holocene marine incursion
(Porter et al., 1984). In the early Holocene, unlike
conditions at Puerto Williams, fire appears to have been
effective in slowing the spread of beech forest. Perhaps
the most striking short-term disruption of vegetation by fire
is centered near 3970 yr BP. This instance, from evidence of
paleo-Indian remains and charcoal excavated at nearby Bahía
Buena and Punta Santa Ana (Ortiz-Troncoso, 1979), is probably
accounted for by human use of fire, which dates from 5895 yr
BP. Instances after 2000 yr BP, effecting less obvious change,
may be traceable not only to paleo-Indians but also to
Spanish settlement which occurred in 1584 A.D. at Ciudad del
Rey Don Felipe, now in ruins adjacent to the mire at Puerto
del Hambre.

2 TORRES DEL PAINE SHORT SECTION

For a detailed view of fire history in the context of
vegetation change during recent centuries, including the time
of settlement, attention is directed to the charcoal profile
for the short section taken of the mire in Parque Nacional
Torres del Paine (Figure 1). Local vegetation classified as
matorral (Pisano, 1974), contains elements both xerophytic
(**Mulinum spinosum**) and mesophytic (**Escallonia rubrum**),
interspersed with groves and lone individuals of young
colonizing deciduous beech. Fire at some time in the recent
past and subsequent erosion are apparent in the park. Numerous
charred trunks of standing and toppled trees, as much as 60 cm
in diameter, represent the remains of preexisting forest.
   Charcoal in the mire, shown at cm intervals in relation to
the pollen and spore stratigraphy (Figure 3), is indicative of
the changing impact of fire on the vegetation. Amounts traced
through five pollen assemblage zones (Table 2) follow
inversely trends in profiles of **Nothofagus** (**N. dombeyi** type;
Heusser, 1971) and other tree taxa. Charcoal is least
abundant in zones TDP-1 and TDP-5 when percentages of
Nothofagus are highest; in zone TDP-3 corresponding with
moderate representation by **Nothofagus**, particles increase and
in zones TDP-2 and TDP-4 are most abundant at levels preceding
decline of **Nothofagus**.

103

Table 2

Pollen assemblages zones in relation to charcoal and age in the short section at Torres del Paine[a]

Zones	Assemblages	Charcoal $(\times 10^2\ cm^{-3})$	Age (yr BP)
TDP-1	**Nothofagus-Empetrum**	1	0-50
TDP-2	Gramineae-**Rumex-Plantago** Liguliflorae-**Nothofagus**	4 5	50-100
TDP-3	**Nothofagus-Embothrium-Maytenus**-Caryophyllaceae-**Misodendrum**	3.2	100-200
TDP-4	Caryophyllaceae-**Acaena**-Tubuliflorae	5.1	200-600
TDP-5	**Nothofagus-Misodendrum**	1	600-660

[a] Charcoal values are means of measurements at sample levels in each zone. Ages of zonal boundaries are estimated from the radiocarbon date of 660 $\pm$ 100 yr BP at the base of TDP-5, from the time of the cold climatic phase implied by late Neoglacial advances of glaciers in the Southern Andes (Mercer, 1982), which is believed to be coincident with the high percentages of Caryophyllaceae in TDP-4 and TDP-3, and from the beginning of settlement a century ago (Martinić, 1974), placed at the TDP-3 - TDP-2 boundary, when taxa of European origin first appear.

Peak charcoal at 10 cm depth is taken to be the marker for fires set at the beginning of the era of European exploration and settlement, which opened a century ago in the area of Parque Nacional Torres del Paine (Martinić, 1974). It is also at this level that adventitious plants, introduced and naturalized from the Old World (Moore, 1983), first become apparent, for example, **Rumex** (cf. **R. acetosa, R. acetosella**), **Plantago** (cf **P. major, P. lanceolata**), and Liguliflorae (cf **Hypochoeria radicata, Taraxacum officinale**). These weedy species, spreading on disturbed ground after burning, register percentage increases in zone TDP-2, and later, because of their competitive inability, decrease in zone TDP-1. Over the past estimated 50 years, as the incidence of fire has decreased, they have been supplanted to a great extent by the succession of **Nothofagus** and other native species. Protection of the area from fire and exploitation took place with establishment of the national park in 1959 A.D. (Pisano, 1974).

Overriding the effects of fire and man, the influence of changing climate of the Little Ice Age is also possibly expressed in the pollen stratigraphy (Figure 3). The interval of the Little Ice Age of the 16th-19th centuries (Lamb, 1977), when temperatures were lower than today, is recorded in the Southern Andes by late Neoglacial advances of glaciers in the Patagonian icefields dating from before 1600 until about 1850 A.D. (Mercer, 1982). Increase of Caryophyllaceae especially in upper zone TDP-4 but also in zone TDP-3 may be explained by this colder climatic regime. In these zones, when percentages of **Nothofagus** and corresponding levels of charcoal are low, open communities are suggested, brought about by a factor other than fire. Caryophyllaceae, as much as 82% of the pollen sum, are provisionally identified as **Cerastium arvense** which grows under open conditions at Torres del Paine in a variety of plant communities between the matorral and the Andean tundra at 1100 m (Pisano, 1974; Moore, 1983). The fossil pollen, identifiable from exine sculpturing and texture, pore number, and size as **Cerastium arvense** (Heusser, 1971), differs on the basis of these morphological criteria from other regional native species, **Arenaria serpens, Colobanthus quitensis, C. subulatus,** and **Silene magellanica** (Moore, 1983). Openness in communities of **Nothofagus,** caused by unfavourable climate when tree line apparently was lower during the Little Ice Age, provided a suitable habitat in which heliophytic Caryophyllaceae could thrive.

ACKNOWLEDGEMENTS

These studies were supported by National Science Foundation grants AMT-7817048, AMT-8115551, and AMT-8308021. I thank L.E. Heusser and S.C. and A.H. Porter for field assistance, the Santiago and Punta Arena offices of the Empresa Nacional del Petróleo (ENAP) for arranging field transportation, especially C. Mordojovich K., E. González P., S. Harambour, and V. Pérez D., and the Chilean Government for the permission to work and collect in Parque Nacional Torres del Paine.

105

REFERENCES

Auer, V. 1950. Las capas volcánicas como base de la cronología postglacial de Fuegopatagonia. **Revista de Investigaciones Agrícolas**, 3:51-208. Buenos Aires.

Auer, V. 1958. The Pleistocene of Fuego-Patagonia. Part II. The history of the flora and vegetation. **Annales Academiae Scientiarum Fennicae**, III, Geologica-Geographica, 50:1-239. Helsinki.

Bird, J. 1938. Antiquity and migrations of the early inhabitants of Patagonia. **Geographical Review**, 28:250-275.

Bridges, E.L. 1948. **The Uttermost Part of the Earth**. Hodder and Stoughton, London.

Cardich, A. 1978. Recent excavations at Lauricocha (central Andes) and Los Toldos (Patagonia). **In: "Early Man in America"**, A.L. Bryan (ed.), p. 296-300. Occasional Papers N°1, Department of Anthropology, University of Alberta.

Cardich, A., Cardich, L.A., & Hajduk, A. 1973. Secuencia arqueológica y cronología radiocarbónica de la Cueva 3 de Los Toldos (Santa Cruz, Argentina). **Relaciones de la Sociedad Argentina de Antropología**, 7:85-123. Buenos Aires.

Casertano, L. 1963. General characteristics of active Andean volcanoes and a summary of their activities during recent centuries. **Seismologica Society of America Bulletin**, 53:1415-1433.

Clark, R.L. 1984. Effects on charcoal of pollen preparation procedures. **Pollen et Spores**, 26:559-576.

Darwin, C. 1958. **The Voyage of the Beagle**. Bantam, New York.

Davis, R.B. 1967. Pollen studies of near-surface sediments in Maine lakes. **In: "Quaternary Paleoecology"**, E.J. Cushing and H.E. Wright, Jr. (eds.), p.143-173. Yale University Press, New Haven.

Deevey, Jr., E.S., Gralenski, L.J., & Hoffren, V. 1959. Yale Natural Radiocarbon Measurements IV. **American Journal of Science, Radiocarbon Supplement**, 1:144-172.

Godley, E.J. 1965. Botany of the Southern Zone. Exploration to 1843. **Tuatara**, 13:140-181.

Grosse I., J.A. 1974. **Visión de Aisén**. Grosse I., Santiago.

Heusser, C.J. 1971. **Pollen and Spores of Chile**. University of Arizona Press, Tucson.

Heusser, C.J. 1983. Quaternary pollen record from Laguna de Tagua Tagua, Chile. **Science**, 219:1429-1432.

Heusser, C.J. 1984a. Late Quaternary climates of Chile. **In: "Late Cainozoic Palaeoclimates of the Southern Hemisphere"**,

J.C. Vogel (ed.), p.59-83. Rotterdam: Balkema.

Heusser, C.J. 1984b. Late-glacial-Holocene climate of the lake district of Chile. **Quaternary Research**, 22:77-90.

Heusser, L.E. & Stock, C.E. 1984. Preparation techniques for concentrating pollen from marine sediments and other sediments with low pollen density. **Palynology**, 8:225-227.

Lamb, H.H. 1977. **Climate: Present, Past, and Future**. Vol. 2. Methuen, London.

Macphail, M.K. 1984. Small-scale dynamics in an early Holocene wet sclerophyll forest in Tasmania. **New Phytologist**, 96:131-147.

Markgraf, V. 1983. Late and postglacial vegetational and paleoclimatic changes in subantarctic, temperate, and arid environments in Argentina. **Palynology**, 7:43-70.

Martinić, B.M. 1973. Panorama de la colonización en Tierra del Fuego entre 1881 y 1900. **Anales del Instituto de la Patagonia**, 4:5-69. Punta Arenas.

Martinić, B.M. 1974. Reconocimiento geográfico y colonización de Ultima Esperanza, 1870-1910. **Anales del Instituto de la Patagonia**, 5:5-53. Punta Arenas.

Martinić, B.M. 1976. La expansión económica de Punta Arenas sobre los territorios argentinos de la Patagonia y Tierra del Fuego, 1885-1925. **Anales del Instituto de la Patagonia**, 7:5-42. Punta Arenas.

Martinić, B.M. 1977. Ocupación y colonización de la región septentrional del antiguo territorio de Magallanes, entre los paralelos 47° y 49° Sur. **Anales del Instituto de la Patagonia**, 8:5-57. Punta Arenas.

Martinić, B.M. 1978. Exploraciones y colonización en la Región Central Magellánica, 1853-1920. **Anales del Instituto de la Patagonia**, 9:5-42. Punta Arenas.

Mercer, J.H. 1982. Holocene glacier variations in southern South America. **Striae**, 18:35-40.

Miller, A. 1976. The climate of Chile. **In: "World Survey of Climatology"**, Vol. 12, "Climates of Central and South America", W.Schwerdtfeger (ed.), p.113-145. Amsterdam, Elsevier.

Montané, J. 1968. Paleo-Indian remains from Laguna de Tagua Tagua, Central Chile. **Science**, 161:1137-1138.

Moore, D.M. 1983. **Flora of Tierra del Fuego**. Nelson, Oswestry.

Ortiz-Troncoso, O.M. 1979. Punta Santa Ana et Bahía Buena: deux gisements sur une ancienne ligne de rivage dans le Détroit de Magellan. **Siege de la Société Musée de l'Homme**, 66:133-204.

Payette, S. & Gagnon, R. 1985. Late Holocene deforestation and tree regeneration in the forest-tundra of Quebec. **Nature,** 313:570-572.

Pisano, V.E. 1973. Fitogeografía de la Península Brunswick, Magallanes. I. Comunidades meso-higromórficas e higromórficas. **Anales del Instituto de la Patagonia,** 4:141-206. Punta Arenas.

Pisano, V.E. 1974. Estudio ecológico de la región continental sur del área andino-patagónica. II. Contribución a la fito- geografía de la zona del Parque Nacional "Torres del Paine". **Anales del Instituto de la Patagonia,** 5:59-104. Punta Arenas.

Pisano, V.E. 1977. Fitogeografía de Fuego-Patagonia chilena. I. Comunidades vegetales entre las latitudes 52° y 56° S. **Anales del Instituto de la Patagonia,** 8:121-250. Punta Arenas.

Pisano, V.E. 1981. Bosquejo fitogeográfico de Fuego-Patagonia. **Anales del Instituto de la Patagonia,** 12:159-171. Punta Arenas.

Porter, S.C., Stuiver, M. & Heusser, C.J. 1984. Holocene sea- level changes along the Strait of Magellan and Beagle Channel, southernmost South America. **Quaternary Research,** 22:59-67.

Prohaska, F. 1976. The climate of Argentina, Paraguay, and Uruguay. In: **"World Survey of Climatology",** Vol. 12, "Climates of Central and South America", W. Schwerdtfeger (ed.), p.13- 112. Amsterdam: Elsevier.

Sanford, Jr., R.L., Saldarriaga, J., Clark, K.E., Uhl, C. & Herrera, R. 1985. Amazon rain-forest fires. **Science,** 227:53- 55.

Singh, G., Kershaw, A.P. & Clark, R.L. 1981. Quaternary vegetation and fire history in Australia. In: **"Fire and Australian Biota",** A.M. Gill, R.A. Groves, and I.R. Nobles (eds.), p.23-54. Australian Academy of Science, Canberra.

Skewes, V.; M.A. 1978. Geología, petrología, quimismo y origen de los volcanes del área de Pali-Aike, Magallanes, Chile. **Anales del Instituto de la Patagonia,** 9:95-106. Punta Arenas.

Swain, A.M. 1973. A history of fire and vegetation in north- eastern Minnesota as recorded in lake sediments. **Quaternary Research,** 3:383-396.

Swain, A.M. 1978. Environmental changes during the past 2000 years in north-central Wisconsin: analysis of pollen, charcoal, and seeds from varved lake sediments. **Quaternary Research,** 10:55-68.

Wright, H.A. & Bailey, A.W. 1982. **Fire Ecology.** Wiley, New York.

Zamora, M.E. 1975. La evolución urbana de Punta Arenas. Creci- miento entre 1848 y 1975. **Anales del Instituto de la Pata-**

gonia, 6:61-92. Punta Arenas.

Zamora, M.E. & Santana, A.A. 1979. Características climáticas de la costa occidental de la Patagonia entre las latitudes 46°40' y 56°30'S. **Anales del Instituto de la Patagonia,** 10:109-144. Punta Arenas.

LUIS SPALLETTI & SERGIO MATHEOS
Centro de Investigaciones Geológicas, CONICET, La Plata, Argentina

DANIEL POIRÉ
Departamento de Geología, Universidad Nacional de Río Cuarto, Argentina

8

Sedimentology of Holocene littoral ridges of Bahía Samborombón (Buenos Aires Province, Argentina)

ABSTRACT

The easternmost Holocene littoral ridge of Bahía Samborombón area (Buenos Aires province, Argentina) shows several sedimentary facies defined by their primary sedimentary structures.

Subhorizontal stratified facies represents beach and surf zone high energy deposits of an open marine wave influenced coast. The intercalations of isolated tabular-planar cross-beds were formed by surf-induced sand waves. Hummocky facies are interpreted as the deposits of breaking-wave bars which evolved in the nearshore environment, very close to the beach. Lenticular cross-bedded facies was formed by 3D dunes oriented parallel to the strandline, while the trough facies is thought to be the result of channeled rip and backwash currents. Amalgamated trough layers were probably formed in intertidal swales.

Six textural types have been recognized in the littoral ridge, ranging from shelly gravels up to fine siliciclastic sand, being the mixed (shell-sand) textural types the more frequent. A convex pattern characterizes the cumulative distribution of the coarse carbonates, while a concave one is typical for the fine sand. The obtained values for the statistical parameters are diverse. Pure textural types tend to be well-sorted and symmetrical, whereas the mixed sediments are poorly sorted and asymmetrical. The textural diversity and the convex-concave pattern of the cumulative distributions suggest that these sediments are the result of several combined littoral and nearshore processes.

The composition and roundness of the heavy and light minerals

of the fine sand population are similar to those of the
Atlantic beach sands of Argentina. The main provenance
(pyroclastic-volcanic) areas were those of the central Argentine
pampas and the Patagonia. Clastic materials were carried
alongshore from the Patagonian Atlantic region by westerly
winds.

RESUMEN

El cordón litoral más oriental de la Bahía Samborombón (provincia de Buenos Aires, Argentina) muestra varias facies sedimentarias definidas a partir de sus estructuras sedimentarias
primarias.

Las facies estratificadas subhorizontales representan depósitos de zonas de playa y rompiente de alta energía de una costa
marina abierta con influencia de ola. Las intercalaciones de
estratos cruzados tabular-planares aislados fueron formadas
por ondas de arena inducidas por la rompiente. Las facies en
montículos son interpretadas como los depósitos de barras de
rompiente las cuales evolucionaron en el ambiente próximo a
la costa, muy cerca de la playa. Las facies de capas cruzadas
lenticulares fueron formadas por dunas 3D orientadas paralela-
mente a la línea de costa, mientras que las facies de artesa
se interpretan como el resultado de corrientes de resaca cana-
lizadas. Las capas en artesa amalgamadas fueron tomadas probablemente en ondas intertidales.

Seis tipos texturales han sido reconocidos en el cordón litoral, variando desde gravas calcáreas a arenas siliciclásticas
finas, siendo los tipos texturales mixtos (arena-conchilla) los
más frecuentes. Un diseño convexo caracteriza la distribución
acumulativa de los carbonatos más gruesos, mientras que uno
cóncavo es típico para la arena fina. Los valores obtenidos
para los parámetros estadísticos son variados. Los tipos texturales puros tienden a ser bien seleccionados y simétricos,
mientras que los sedimentos mixtos son pobremente seleccionados
y asimétricos. La diversidad textural y el diseño convexo-cón-
cavo de las distribuciones acumulativas sugieren que estos
sedimentos son el resultado de varios procesos litorales y costeros proximales combinados.

La composición y redondez de los minerales livianos y pesados
de la población de arena fina son similares a aquellas de las
arenas de las playas atlánticas argentinas. Las principales

áreas de proveniencia piroclástica-volcánica fueron las pampas
de Argentina central y la Patagonia. Los materiales clásticos
fueron llevados a lo largo de la costa desde la región patagónica atlántica por los vientos occidentales.

INTRODUCTION

Several Holocene littoral ridges with their major axes
parallel to the coastline are placed in the easternmost Buenos
Aires Province, Argentina.

The deposits of these ridges were clearly described by
Frenguelli (1950), Cappannini (1952) and Tricart (1968). In
the study area of the Bahía Samborombón, this unit is known
as the Cerro de la Gloria Member of Las Escobas Formation
(Fidalgo et al., 1973). Further south, it was named Los Zorza-
les Formation (Dangavs, 1983) and Mar Chiquita Formation
(Schnack et al., 1982).

The ridges were developed on a fine grained lowland formed
during the Querandinense ingression (Frenguelli, 1950), though
they can also be placed over other Quaternary units (Fidalgo
et al., 1973). Radiocarbon dating of articulated mollusc shells
resting in life position yielded ages of between 3850 ± 60
and 1340 ± 50 years BP for Mar Chiquita Formation (Schnack
et al., 1982). Based on ^{14}C data, Gómez et al. (1985-1987)
suggested that the Cerro de la Gloria Member is younger than
4440 ± 110 years BP.

This paper deals with the sedimentology of the easternmost
Holocene ridge placed in the nearabouts of Río Samborombón
and Río Salado mouths (Figure 1). In order to determine the
environments and the processes of accumulation we defined
several sedimentary facies, based on the features and
orientation of primary sedimentary structures. The grain size
distribution and the mineral composition of the deposits were
also studied and evaluated.

GENERAL FEATURES OF THE DEPOSITS

The Cerro de la Gloria Member comprises a well-stratified
succession of white fine to coarse-grained shelly gravels and
light brown fine-grained siliciclastic sands.

The gravel fraction is mainly composed of shells, where

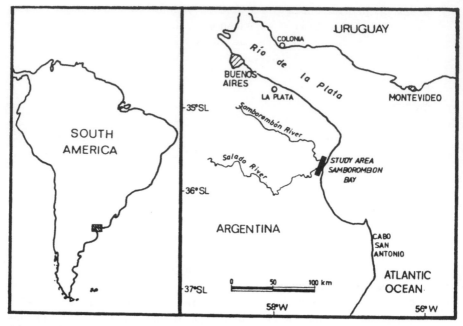

Figure 1. Location map. Studied area in black.

Mactra isabelleana valves prevail, followed by rounded and sub-rounded pebbles of caliche. The gravels appear as plane-bedded and cross-bedded sedimentary units, 5 to 40 cm thick. They sometimes show normal graded stratification.

The terrigenous sandy deposits are well sorted. They occassionally include a few broken shells and vertebrate bones and teeth. Their beds are frequently 1 to 8 cm thick, though they can reach 25 cm.

SEDIMENTARY FACIES

The deposits of the Holocene coastal ridges are formed by four sedimentary facies, clearly recognized by the morphology of their bodies and by their primary sedimentary structures (Figure 2). They are: 1) subhorizontal stratified facies, 2) hummocky facies, 3) lenticular (planar to tangential) cross-bedded facies, and 4) trough facies.

1. SUBHORIZONTAL STRATIFIED FACIES

The most typical features of the ridges are more than 1 meter
thick sequences of 6° to 14° ESE dipping strata. The facies
is characterized by varied grain-sized shelly gravels and fine
sand beds, which are well-defined by sharp changes in grain
size. The sandy and fine-grained gravel deposits are 1 to 6 m
thick and the medium-grained (pebbly) gravels are between 10
and 40 cm thick.

All the sediments of this facies are well sorted. Despite
the differences in shape, there is a remarkable similitude in
grain size between the shell fragments and the caliche clasts.
The shells are differently oriented, but they frequently show
low-angle imbrications towards the SE.

Tabular-planar (and tangential) cross-beds can be found as
isolated sets within the subhorizontal stratified facies, or
forming a marked alternating sequence with these beds. The
cross-bedded units, 5 to 40 cm thick, are composed of thin,
sometimes graded, shelly gravels and snady foresets dipping
20° NW. Some of these beds show a microdelta pattern, with
progressive increase of the foreset dip towards the direction
of accretion.

2. HUMMOCKY FACIES

This facies is characterized by 1 to 1.5 m thick sequences
formed by 2 to 2.5 m wide hummocks (Figure 2), showing a
lateral continuity of more than 100 m.

Their deposits consist of alternating shelly gravels and
sands. There is a good positive correlation between bed
thickness and grain size, though lateral variations in the
thickness of individual layers are observed passing from the
antiforms to the sinforms of the hummocks.

Coarse-grained gravel levels are more frequent towards the
upper part of the hummocks. Poorly-sorted gravels with
disordered shelly clasts (some of them in vertical position)
were deposited on the crest of these features. The better
sorted gravels show some imbrication of both the shells and
the caliche clasts. In some levels the biogenic clasts are
found in horizontal position with their convex face upwards.

The hummocky pattern is similar in every direction. In the
NW-SE sections it is shown that these sequences were deposited
on a regional paleoslope inclined 6° to 10° SE.

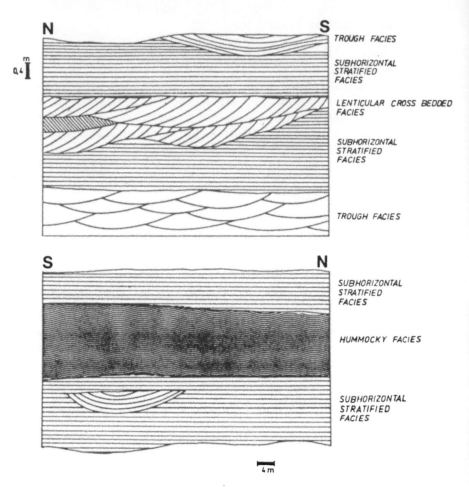

Figure 2. Schematic representation of facies distribution in the Holocene littoral ridge.

3. LENTICULAR CROSS-BEDDED FACIES

A laterally discontinuous cross-bedded coset of concave base and flat to wavy (bar-like) top characterizes this facies (Figure 2). Every cross-bedded set varies laterally in thickness between 10 and 50 cm.

Each set is bounded by second order surfaces. The tangential and planar foresets are less than 5 cm thick, and they are seldom cut by reactivation surfaces. The majority of the cross-sets and reactivation surfaces are oriented northwards, though some isolated beds have an opposite orientation.

As in the subhorizontal stratified facies, the sediments are well-sorted white gravels and light brown siliciclastic sands.

4. TROUGH FACIES

It is composed of shelly gravel and sand sequences of troughs with their axes oriented NW-SE.

Two varieties can be distinguished according to a scale:
a) medium scale troughs, between 50 to 70 cm thick and 60 to 80 cm wide, and
b) large scale troughs, more than 60 cm thick and over 80 cm wide.

This facies is typically found as sequences of lateral and vertical lens-shaped troughs. However, sometimes it is found as isolated sets intercalated in the subhorizontal stratified facies (Figure 2).

5. INTERPRETATION OF FACIES

It is believed that each facies represents a typical depositional feature of the littoral and the sub-littoral environments.

Subhorizontal stratified bodies are interpreted as beach and surf zone deposits of an open marine coast washed by waves. The inclination angle of the accretion bedding suggests a high energy nearshore environment. Sediments represent processes of strong vertical accretion rate under conditions of both high current velocity and velocity asymmetry (Clifton, 1976). The disponibility of detrital material was high, especially of shelly and caliche clasts.

The tabular cross-beds intercalated in the subhorizontal strata would represent surf induced sand waves located between the outermost breakers and the beach, since their paleocurrents are completely opposed to the paleoslope. We think that these beds were formed by the migration of low flow regime sand waves during periods of decreasing vertical accretion rate.

Sections with interstratifications of accretion beds and tabular cross beds are probably reflecting slight variations in sea level and/or cyclic changes in the kynetic energy of waves and related currents.

Hummocky facies shows events of strong energy release. The
upward-coarsening tendency and the random orientation of shells
in the higher parts of the deposits reveal a marked increase
of turbulence towards the top of the bars. Textures and primary
structures suggest the action of breaking waves, perhaps
during storms, seiching (Duke, 1985) or surging waves (Galvin,
1968; Spalletti, 1986). Their development over vertical
accretion layers reveals that these bars evolved in the
nearshore environment close to the strandline.

The lenticular cross-bedded facies was probably formed by
the migration of tridimensional bars (dunes) induced by
longitudinal currents, as it is suggested by the geometry of
the bodies and the orientation of foresets. This subtidal-
intertidal deposit is considered to have been generated by
fast sub-critical currents (Clifton, 1976; Boothroyd, 1982;
Bohacs and Southard, 1982) oriented parallel to the beach. The
abundance of reactivation surfaces suggests destructive events
reflecting periods of high wave energy and/or low tide. The
northward paleocurrent trends indicate longshore drift very
similar to the present one in the sub-littoral zone of the
Argentinian Atlantic ocean (Mazzoni and Spalletti, 1978).

The trough facies was formed by cut-and-fill processes
perpendicular to the strandline. We think that flow conditions
were in the uppermost lower flow regime. Isolated trough beds
intercalated in subhorizontal stratified facies were probably
formed by channelled wave-related rip and backwash currents.
The levels of lateral and vertical amalgamated trough sets are
here interpreted as the deposits of intertidal swales.

GRAIN SIZE ANALYSIS

The sediments were sieved following Ingram's method (1971) at
intervals of 0.5 $\emptyset$, from -3.5 $\emptyset$ up to 4 $\emptyset$. The results of the
mechanical analysis have been represented in histograms and
cumulative probabilistic diagrams (Figure 3). Folk and Ward's
(1957) and Passega's (1957) statistical parameters were also
calculated (Table 1).

The studied deposits are composed by siliciclastic sands and
by carbonates, the latter being formed by both shelly
fragments and, less abundant, caliche calcilites. Thus, the
grain size determinations were done thoroughly, including
the carbonates, as recommended by Passega (1957) and Shepard
and Young (1961).

118

Six textural types have been recognized considering the proportions of the siliciclastic and carbonate end members and the grain size of the carbonate material: 1) shelly gravel, 2) shelly gravel with fine siliciclastic sand, 3) very-coarse shelly sand with fine siliciclastic sand, 4) very-coarse shelly sand, 5) fine siliciclastic sand with very-coarse shelly sand, and 6) fine siliciclastic sand.

One of the most remarkable features of the frequency distribution is the emplacement of the siliciclastic within the fine sand class interval (3 $\emptyset$), whereas the carbonates appear in different fractions, from -0.5 up to -3.5 $\emptyset$.

The pure textural types show marked modes, whereas there is a conspicuous bimodal (sometimes polymodal) distribution pattern in the mixed textural types, with slightly marked modes (Figure 3). In the grain size frequency distribution of all mixed samples, the siliciclastic and carbonate end members are easily recognizable. This feature is more noticeable for coarser carbonate debris.

The cumulative distribution diagrams show characteristic patterns for each textural type. A convex pattern with higher sorted coarse segment is typical for the pure carbonate sediments (textural types 1 and 4, figure 3). The pure siliciclastic sands (textural type 6, figure 3) have cumulative distributions with a well-sorted concave pattern, very similar to that of modern forebeach sands of Cabo San Antonio (Spalletti and Mazzoni, 1979).

In the mixed sediments, there exists a combination of coarse convex population and a fine one with well-sorted concave pattern (Figure 3).

The mean (Mz $\emptyset$), median ($\emptyset$ 50) and 1 percentile (C $\emptyset$) values show the wide range of grain sizes found in the studied deposits (Table 1). Some deficiencies in the medium and coarse sand classes are seen when analyzing 0 50 values.

The sorting, as measured by 0 standard deviations, is variable (Table 1), though poorly sorted sediments predominate. As expected, the pure textural types are the best sorted sediments.

Mean and standard deviation values of the fine siliciclastic pure sand are comparable with those of the Cabo San Antonio beach sands (Spalletti and Mazzoni, 1979).

A wide range of skewness (Sk1) values was obtained for the different textural types (Table 1). Pure sediments tend to be symmetrically distributed. The mixtures with predominance of

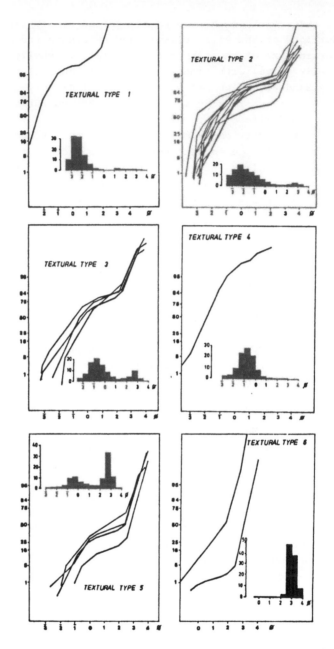

Figure 3. Log probability plots of cumulative grain size distributions and representative histograms of the six textural types.

the coarse end member are positively skewed whereas the fine sand rich mixtures are negatively skewed.

Kurtosis values are also highly varied. There is a tendency to platikurtic distributions in sandy pure sediments (Table 1).

The scatter plot of Figure 4 shows the bi-variable relationship between Mz and σ_1 . Though the studied samples are quite coarser than the ones studied by the authors, most of them fall in the fluvial-beach superimposed field recognized by Mazzoni (1977b).

CM diagram (Figure 4) shows a close positive correlation between ϕ 1 and ϕ 50 values, and the differencies between both parameters are more conspicuous towards the finer sediments.

1. INTERPRETATION

Most textural studies in littoral environments were carried out on sandy sediments (Mason and Folk, 1958; Shepard and Young, 1961; Friedman, 1961; Hails, 1967; Klovan, 1966; Glaister and Nelson, 1974; Mazzoni, 1977b; Spalletti and Mazzoni, 1979; Kennedy et al., 1981). Kumar (1973) analyzed some textural mixtures in bar sediments, whereas Dyer (1970a, 1970b) dealt with sandy gravel deposits of marine environment. The characteristic bimodal distribution and complex cumulative frequency patterns of mixed textural types are very similar to Dyer's data.

We considered that the grain size distributions described above are typical for high energy open shallow marine and coastal sediments. The depositional environment was characterized by abundant supply of carbonate skeletal material and lack of coarse grained siliciclastic sands.

The textural diversity and the complex convex-concave pattern of the cumulative distributions do not allow any definition of the traditional traction, saltation and suspension population suggested by Moss (1962) and Visher (1969). Several processes are believed to have occurred during the building of the ridges, such as upward displacement of debris during breakers, suspension transport by surf currents and traction-saltation transport by backwash and/or rip currents, as well as by longitudinal currents. Thus, the grain size distributions of the studied sediments are the result of diverse flow conditions and of a complex pattern of currents.

The competence of the depositional agents is clearly seen by the grain size of the carbonate population, because the

Table 1. Grain size parameters of the studied samples.

Sample	Textural type	ϕ 1	ϕ 50	Mz ϕ	$\sigma_{1\phi}$	Sk$_{1\phi}$	Kg ϕ
A 5	5	-1.85	1.30	1.19	1.637	-0.127	0.669
A 4	5	-0.97	2.82	2.53	1.091	-0.533	2.023
A 3	3	-2.40	-0.46	0.16	1.579	0.481	1.227
A 3bis	2	-2.70	-2.00	-1.58	1.376	0.640	1.374
A 2	2	-3.30	-2.12	-1.84	1.443	0.460	1.086
A 1	3	-3.15	-0.65	-0.06	1.938	0.341	0.476
A-1	6	0.27	3.00	3.03	0.338	0.124	0.919
M 3	2	-3.80	-3.09	-2.44	1.672	0.768	2.183
M 2	2	-3.00	-1.66	-1.39	1.301	0.507	2.208
M 1	6	-1.05	1.90	1.65	0.877	-0.439	0.957
M-1	5	-1.71	2.60	1.69	1.569	-0.708	0.635
B12	2	-2.90	-1.00	-0.90	1.137	0.172	1.426
B11	2	-3.22	-1.72	-1.55	1.461	0.354	1.337
B10	3	-1.70	0.12	0.39	1.328	0.260	0.978
B 9	4	-2.65	-0.82	-0.81	0.733	0.016	1.102
B 8	2	-2.76	-1.70	-1.56	1.218	0.402	1.299
B 7	1	-3.25	-2.40	-2.35	0.590	0.215	1.170
B 6	2	-2.65	-0.98	-0.83	1.274	0.328	1.836
B 5	2	-3.16	-1.35	-1.12	1.338	0.378	1.681
B 4	2	-3.10	-0.95	-0.21	2.103	0.407	0.560
B 3	5	-2.35	2.30	1.60	1.532	-0.587	0.635
B 2	3	-3.15	-0.30	0.19	1.790	0.305	0.791
B 1	2	-3.23	-1.25	-0.96	1.676	0.323	1.144

siliciclastic materials only contribute to the finer sand
grades.

Therefore, the good sorting of the pure texture types shows
the high selectivity of the littoral currents. Consequently,
the mixed sediments are thought to be formed by infiltration
of the siliciclastic sands in the pores of the carbonate gravels
and coarse carbonatic sands.

COMPOSITION AND ROUNDNESS OF THE SILICICLASTIC POPULATION

The mineral composition was determined in the 88-125 class of
11 selected samples (Table 2). The study consisted in the
separation of heavy and light minerals and their microscopic
analysis. Counting was done following Carver's (1971) method

and the roundness was calculated according to Power's (1953) visual scale.

The content of heavy minerals varies between 5.4% and 18.4%, with an average of 10.7%. These data are similar to the ones obtained by Teruggi et al. (1959) in sands of the coastal area of Buenos Aires Province, though they are higher than the ones of the modern fore-beaches of Cabo San Antonio (Mazzoni, 1977).

The light mineral association is feldspar (plagioclase and K-feldspar), quartz, lithic fragments and a few glass shards.

Plagioclase is one of the main components (29%-45%), and it is presented with diverse degree of alteration in two compositional varieties, labradorite and oligoclase-andesine. The features of the plagioclase grains (twinning, shapes, inclusions) are like those of the modern coastal sands, studied by Teruggi et al. (1959).

Quartz is a less frequent component (6%-12%). It mainly appears as clean grains with normal extinction. Some poly-crystalline grains of chalcedony can also be found.

K-feldspar (mainly altered orthoclase) is less than 7%. Acid glass shards are very scarce and similar to those of the Quaternary Pampean loessic sediments (Teruggi, 1955).

Volcanic rock fragments (VRF) are very common in these sands (38%-50%). Felsic VRF are abundant, followed by andesitic and basaltic clasts. Clay galls coated with a ferrigenous film were also recognized.

The association and features of heavy minerals are similar to those found by Teruggi et al. (1959), Teruggi (1964) and Mazzoni (1977a) in Atlantic littoral sands of Argentina. The main components are pyroxenes (augite hyperstene), VRF, opaque minerals (magnetite, ilmenite, hematite and leucoxene) and amphiboles (green and brown hornblende and lamprobolite) (Table 2). There are also few scattered grains of garnet, epidote, zircon, tourmaline and micas.

The roundness of light components is quite uniform, with an average of 0.44 (sub-rounded). Heavy minerals always appear more rounded than their light equivalents; their average values range from 0.46 up to 0.51, with an overall mean of 0.48 (Table 2). The roundness of the heavies is slightly higher than that found by Mazzoni (1977a) in modern beach sands of Cabo San Antonio, though similar to that of the heavies of Atlantic beaches southward.

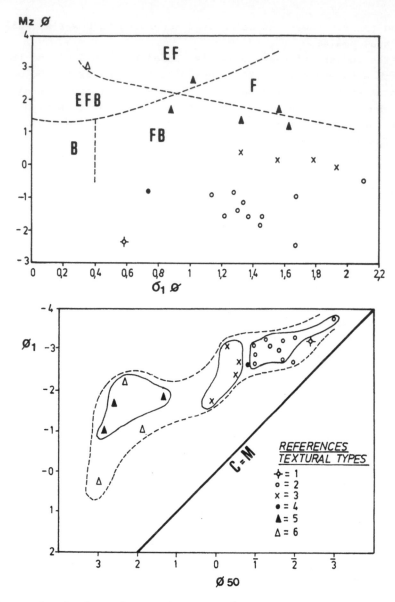

Figure 4. a) Plot of graphic mean (Ø) and graphic standard
deviation (Ø). Environmental fields according to Mazzoni
(1977b). E: aeolian, F: fluvial, B: beach.
b) CM plot.

1. INTERPRETATION

No marked differences were found in composition and roundness
between the several textural types and sedimentary facies.
Only those facies formed under higher energy conditions
(hummocky and subhorizontal stratified facies) have greater
amounts of heavy minerals.

The high proportion and roundness of heavy minerals suggest
that the sand accumulation in the ridges was produced by more
competent mechanisms in comparison with those acting on the
nearshore and beaches of Cabo San Antonio. The slightly lower
roundness of the light minerals can be attributed to their
low density and susceptibility to tractive transport by drift
currents, as well as to some aeolian supply of juvenile
(pyroclastic) materials.

It is evident that studied sands have the same sources as
the present sediments of the Atlantic coast of Buenos Aires
Province. According to Mauriño (1956), Teruggi et al. (1959),
Teruggi (1964) and Mazzoni (1977a), the main source rocks would
be the Tertiary-Quaternary Pampean sediments and those of the
Patagonian Atlantic region, especially of the Río Negro mouth.
There would be some aeolian supply from primary ash falls as
well as by deflation from arid zones of central and southern
Argentina.

QtFL and QmPK diagrams of Figure 5 show a very concentrated
compositional distribution in the field of magmatic arc
provenance (Dickinson and Suczek, 1979; Maynard et al., 1982).
The bulk composition of the sands of the Río de La Plata is
like that of the studied sands, though richer in Qt and Qm
(see Bercowski, 1986). Thus, it is inferred that the influence
of the Río de La Plata source was negligible, as it now occurs
with the modern sands of the coastal zone of Cabo San Antonio
(Urien, 1967; Mazzoni, 1977a).

The heavy-minerals diagram (Figure 5) reinforces the afore-
mentioned conclusions. It clearly shows the abundance of
pyroxenes and amphiboles derived from volcanic terranes, as
compared to the opaque set and the cortical derived translucid
minerals (garnet, apatite and tourmaline).

CONCLUSIONS

1. Four sedimentary facies have been defined in the Holoceno
littoral ridges of Bahía Samborombón.

Table 2. Composition and roundness of siliciclastic sands (88 - 125 interval)

	A-1	A 1	A 4	A 5	B 1	B 4	E6	B11	M-1	M 1	M 3
Total light minerals %	84.43	85.87	91.86	92.20	92.04	87.87	90.53	90.77	89.52	81.59	94.60
Qz %	12.28	11.21	11.82	10.00	7.34	10.58	9.80	11.54	10.28	6.09	8.77
Plg %	38.60	44.85	38.18	40.91	41.28	40.38	29.41	44.23	37.38	40.87	34.21
KF %	3.50	3.74	1.82	2.73	---	0,96	4.90	1,92	---	1.74	6.14
RF %	42.98	38,32	45.45	43.64	45.87	45.19	50.98	38.46	49.53	48.70	46.49
Glass shards %	0.88	1.87	0.91	---	0.92	0.96	0.98	1.92	---	---	0.88
Mean roundness %	0.42	0.41	0.42	0.44	0.45	0.44	0.40	0.45	0.46	0.47	0.45
Total heavy minerals %	15,57	14.13	8.14	7.80	7.96	12.13	9.47	9.23	10.48	18.41	5.40
Clinopx %	23.74	29.81	26.54	28.57	31.54	34.64	32.47	33.33	33.33	37.11	36.18
Hy %	15.11	12.42	14.81	19.29	28.19	15.69	20.13	21.57	25.69	22.01	17.76
Horn. %	10.79	14.29	10.49	11.43	4.70	7.84	3.25	13.07	11.81	6.92	9.87
Lampr. %	2.88	1.24	1.85	---	3.56	2.61	---	1.96	2.08	1.89	1.32
RF %	23.02	26.71	29.01	27.86	17.45	19.61	22.73	13.07	18.75	14.46	17.10
Apatite %	10.07	6.21	4.32	1.43	3.26	1.96	2.60	1.31	2.08	---	3.95
Garnet %	---	---	---	1.43	0.67	---	0.65	2.61	0.69	2.52	1.32
Opaque min. %	14.39	9.32	12.96	10.00	8.72	16.34	14.93	11.11	4.86	11.95	12.50
Epidote %	---	---	---	---	2.01	1.31	3.25	1.96	0.69	3.14	---
Mean roundness %	0.48	0.49	0.46	0.46	0.49	0.48	0.47	0.51	0.50	0.51	0.48

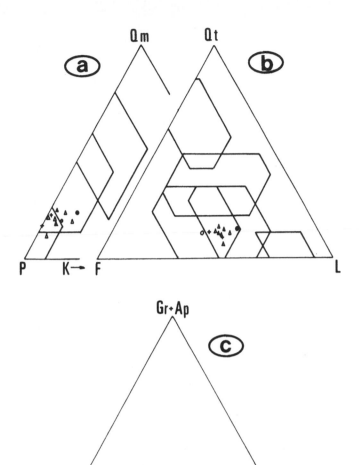

Figure 5. Compositional triangular plots. a) Qm PK plot,
b) QtFl plot and c) garnet plus apatite-opaque-pyroxene plus
amphibole plot.
Subhorizontal facies (triangles), lenticular cross-bedded
facies (full circles), hummocky facies (empty circles) and
trough facies (crosses).

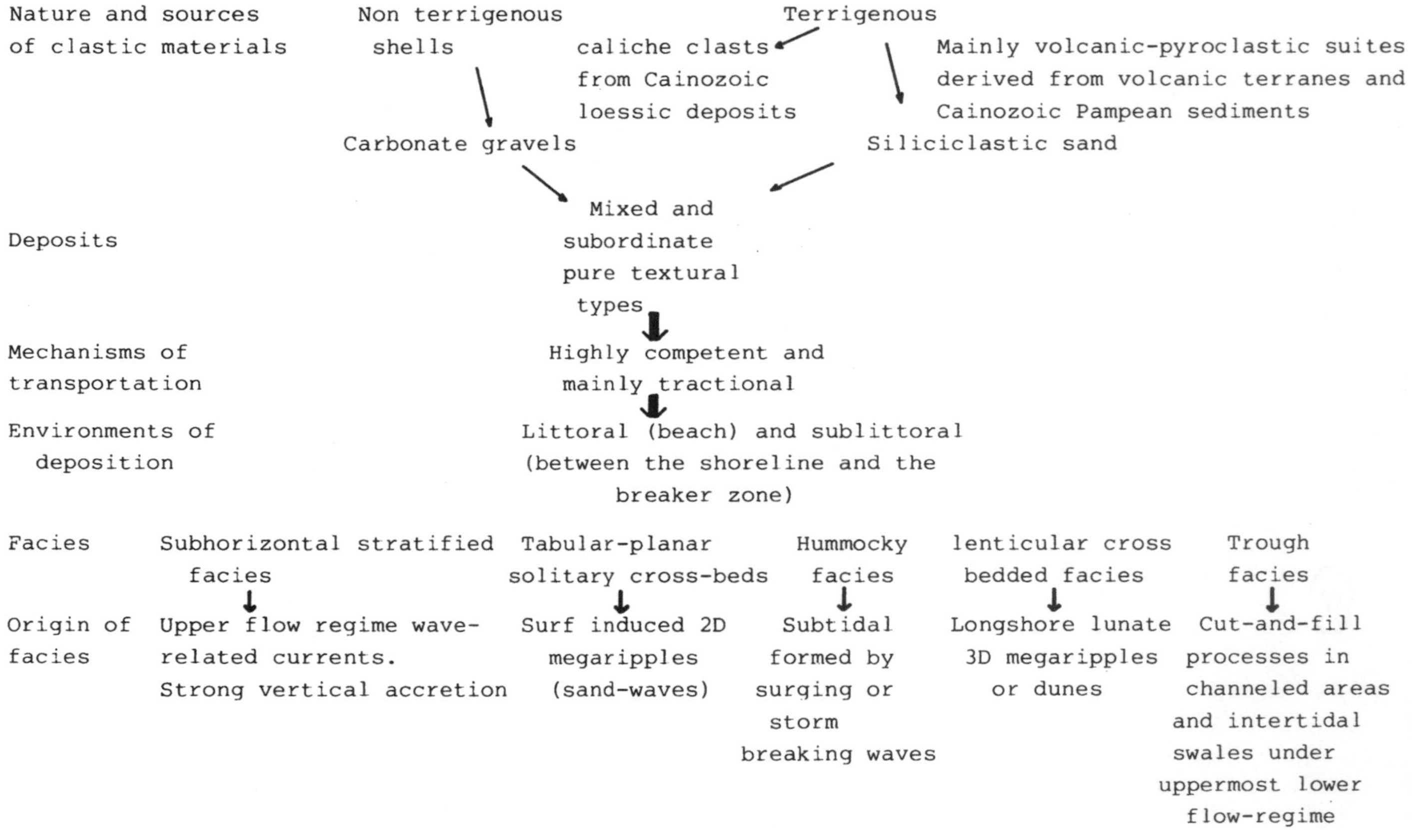

Nature and sources of clastic materials

Non terrigenous
shells
caliche clasts from Cainozoic loessic deposits
Carbonate gravels

Terrigenous
Mainly volcanic-pyroclastic suites derived from volcanic terranes and Cainozoic Pampean sediments
Siliciclastic sand

Deposits
Mixed and subordinate pure textural types

Mechanisms of transportation
Highly competent and mainly tractional

Environments of deposition
Littoral (beach) and sublittoral (between the shoreline and the breaker zone)

Facies
Subhorizontal stratified facies
Tabular-planar solitary cross-beds
Hummocky facies
lenticular cross bedded facies
Trough facies

Origin of facies
Upper flow regime wave-related currents. Strong vertical accretion
Surf induced 2D megaripples (sand-waves)
Subtidal formed by surging or storm breaking waves
Longshore lunate 3D megaripples or dunes
Cut-and-fill processes in channeled areas and intertidal swales under uppermost lower flow-regime

Figure 6. General conceptual model for the studied Holocene littoral ridge.

Subhorizontal stratified facies has been deposited by upper flow regime currents and strong vertical accretion rates. Tabular-planar cross-beds intercalated as isolated sets in these sequences, represent surf-induced 2D sand waves.

Hummocky facies has been formed as subtidal bars accreted by surging and storm waves, while lenticular cross-bedded facies represents the amalgamation of longshore lunate (3D) megaripples. Cut-and-fill processes in channelled areas perpendicular to the shoreline originated the trough facies.
2. Six mixed and pure textural types have been recognized in the littoral ridge based on characteristic patterns of grain-size distribution. Statistical parameters are variable, though poorly-sorted sandy shelly gravels predominate.

The grain size features are considered the result of diverse mechanical processes, such as upward displacement of debris during breakers, tractive (and suspensive) transportation by surf, longshore, swash-backwash and rip currents.
3. There is a complex suite of light and heavy minerals in the fine siliciclastic sand population, but the more abundant are the ones derived from volcanic-pyroclastic terranes. The main sources would be the Cainozoic Pampean sediments and the sands of the Patagonian Atlantic region.

A general conceptual model for the littoral ridges has been formulated (Figure 6) to show the relationships in the observed data.

REFERENCES

Bercowski, F. 1986. Arenas del Río de La Plata: una excepción a la relación entre composición de areniscas y la tectónica de placas. **Res. Expand. 1° Reun. Arg. Sediment.**: 263-266.
Bohacs, K. and Southard, j.B. 1982. Flume studies on the kinematics of large-scale bed forms. **Abs. 11° Int. Congr. Sedim.**: 71.
Boothroyd, J.C. 1982. Mesotidal bedforms revisited: megaripples, sand waves, and transverse bars. **Abs. 11° Int. Congr. Sedim.**: 71.
Cappannini, D. 1952. Geoedafología del curso inferior del Río Salado de la provincia de Buenos Aires. **IDIA**, 50-51, 56pp.
Carver, R.E. 1971. **Procedures in sedimentary petrology.** Wiley Intersc.: 653pp.
Clifton, H.E. 1976. Wave formed sedimentary structures

-a conceptual model. In Davis, R.A. and Ethington, R.L.
(Eds.). **Soc. Econ. Pal. Min. Spec. Publ.** 24:126-148.

Dangavs, N. V. 1983. Geología del Complejo lagunar Salada
Grande de General Lavalle y General Madariaga, provincia
de Buenos Aires. **Asoc. Geol. Arg. Rev.** 38(2):161-174.

Dickinson, W.R. and Suczek, C.A. 1979. Plate tectonics and
sandstones composition. **Am. Assoc. Petrol. Geol. Bull.**
63:2164-2182.

Duke, W.L. 1985. Hummocky cross-stratification, tropical
hurricanes, and intense winter storms. **Sedimentology,**
32:167-194.

Dyer, K.R. 1970. Grain size parameters for sandy gravels.
Jour. Sed. Petrology, 40(2):616-620.

--------- 1970. Sediment distribution in Christchurch Bay,
S. England. **J. Mar. Biol. Ass. U.K.** 50:673-682.

Fidalgo, F.; Colado, U.R. and De Francesco, F.O. 1973. Sobre
ingresiones marinas cuaternarias en los partidos de Castelli,
Chascomús y Magdalena (Pcia de Buenos Aires). **5° Congr. Geol.
Arg.,** 3:227-240.

Folk, R.L. 1966. A review of grain size parameters.
Sedimentology, 6:73-93.

--------- and Ward, W.C. 1957. Brazos River bar: a study in
the significance of grain size parameters. **Jour. Sed. Petrol.**
27(1):3-26.

Friedman, G.M. 1961. Distinction between dune, beach and river
sands from their textural characteristics. **Jour. Sed.
Petrology.,** 31(4):514-522.

--------- 1967. Dynamic processes and statistical parameters
compared for size frequency distribution of beach and river
sands. **Jour. Sed. Petrology.,** 37(2):327-355.

Frenguelli, J. 1950. Rasgos generales de la morfología y de la
geología de la provincia de Buenos Aires. **Publ. LEMIT,**
Bs. As., ser. II, N° 30: 72pp.

Galvin, C.J. 1968. Breaker type classification on three
laboratory beaches. **Jour. Geoph. Res.** 73p. 3651-3659.

Glaister, R.P. and Nelson, H.W. 1974. Grain-size distributiors,
an aid in facies identification. **Bull. Canadian Petrol. Geol.**
22(3):203-240.

Gómez, G.J.; Huarte, R.A.; Figini, A.J.; Carbonari, J.E.;
Zubiaga, A.C. and Fidalgo, F. 1985. Análisis y comparación
de dataciones radiocarbónicas de conchas de moluscos de la
Formación Las Escobas, Provincia de Buenos Aires. **1° Jorn.
Geol. Bonaerenses, Resúmenes:** 121-122.

---------; Figini, A. and Fidalgo, F. 1987. Secuencia vertical de edades ^{14}C en la Formación Las Escobas, en la localidad de Cerro de la Gloria, Bahía de Samborombón, Provincia de Buenos Aires, Argentina. **X Congr. Geol. Arg.** (in press).

Hails, J.R. 1967. Significance of statistical parameters for distinguishing sedimentary environments in New South Wales, Australia. **Jour. Sed. Petrology**, 37(4):1059-1069.

Ingram, R.L. 1971. Sieve Analysis. **In.**: Carver, R. (ed.). **Procedures in Sedimentary Petrology**: 49-67. Wiley.

Kennedy, S.K.; Ehrlich, R. and Kana, T.W. 1981. The non-normal distribution of intermittent suspension sediments below breaking waves. **Jour. Sed. Petrology**, 51(4):1103-1108.

Klovan, J. E. 1966. The use of factor analysis in determining depositional environment from grain size distributions. **Jour. Sed. Petrology**, 36(1):115-125.

Kumar, N. 1973. Modern and ancient barrier sediments: new interpretation based on stratal sequence in inlet-filling sands and on recognition of nearshore storm deposits. **Annals New York Acad. Sci.**, 220(5):245-340.

Mason, C.C. and Folk, R.L. 1958. Differentiation of beach, dune and aeolian flat environments by size analysis, Mustang Island, Texas. **Jour. Sed. Petrology**, 28(2):211-226.

Mauriño, V. 1956. Los sedimentos psamíticos actuales de la región costera comprendida entre Faro Recalada y Faro Monte Hermoso. **Publ. LEMIT** ser. II, 61:1-35.

Mazzoni, M. 1977a. Caracteres composicionales de la fracción pesados de arenas de playa frontal del litoral atlántico bonaerense. **Asoc. Min. Petr. Sed. Rev.** VIII, 3-4:73-91.

---------- 1977b. El uso de medidas estadísticas texturales en el estudio ambiental de arenas. **Obra Centenario Museo La Plata**, IV(Geol.):179-223.

---------- and Spalletti, L.A. 1978. Evaluación de procesos de transporte de arenas litorales bonaerenses a través de análisis texturales y mineralógicos. **Actas Oceanogr. Arg.** 2(1):51-67.

Maynard, J.B.; Valloni, R. and Yu, H.S. 1982. Composition of modern deep sea sands from arc-related basins. **Geol. Soc. London Spec. Publ.** 10:551-561.

Moss, A.J. 1962. The physical nature of common sandy and pebbly deposits. I and II, **Am. Jour. Sci.** 260:337-373; 261:297-343.

Passega, R. 1957. Texture as characteristics of clastic deposition. **Am. Assoc. Petrol. Geol. Bull.** 41:1952-1984.

Powers, M.C. 1953. A new roundness scale for sedimentary particles. **Jour. Sed. Petrology**, 23(1):117-119.

Schnack, E.J.; Fasano, J.L. and Isla, F.I. 1982. The evolution of Mar Chiquita lagoon coast, Buenos Aires Province, Argentina. In: Colquhoun, D. J. (ed.), **Holocene sea level fluctuations, magnitude and causes.** IGCP-INQUA, Columbia, S.C., USA:143-155.

Shepard, F. and Young, R. 1961. Distinguishing between dune and beach sands. **Jour. Sed. Petrology**, 31(2):196-214.

Spalletti, L. 1986. Nociones sobre transporte y depositación de sedimentos clásticos. **Fac. Cienc. Nat. y Museo, Ser. Tecn. y Didact.** 13:102pp. La Plata.

Spalletti, L.A. and Mazzoni, M.M. 1979. Caracteres granulo-métricos de arenas de playa frontal, playa distal y médano del litoral bonaerense. **Asoc. Geol. Arg. Rev.**, 34(1):12-30.

Teruggi, M.E. 1955. Algunas observaciones microscópicas sobre vidrio volcánico y ópalo organógeno en sedimentos pampeanos. **Notas Museo La Plata**, 18(66):17-26.

Teruggi, M.E. 1964. Las arenas de la costa de la provincia de Buenos Aires entre Bahía Blanca y Río Negro. **Publ. LEMIT**, ser. II, 81:38pp.

----------; Chaar, E.; Remiro, J.R. and Limousin, T.A. 1959. Las arenas de la costa de la provincia de Buenos Aires entre Cabo San Antonio y Bahía Blanca. **Publ. LEMIT**, ser, II, 77:37pp.

Tricart, J.L.F. 1968. La geomorfología de la Pampa Deprimida como base para los estudios edafológicos y agronómicos. **Publ. INTA**:138pp.

Urien, C.M. 1967. Los sedimentos modernos del Río de La Plata éxterior. **Bol. Serv. Hidrogr. Naval**, 4, 2. Buenos Aires.

Visher, G.S. 1969. Grain size distributions and depositional processes. **Jour. Sed. Petrology**, 39(3):1074-1106.

LUIS A.ORQUERA
Asociación de Investigaciones Antropológicas, Buenos Aires, CONICET, Argentina

ERNESTO L.PIANA
Centro Austral de Investigaciones Científicas, Ushuaia, Tierra del Fuego, CONICET, Argentina

9

Human littoral adaptation in the Beagle Channel region: The maximum possible age

ABSTRACT

In Tierra del Fuego two contrasting adaptive patterns coexisted: pedestrian hunters and gatherers in the Isla Grande and sea-littoral adapted canoemen in the channels and island coasts. Maximum known ages and possible peopling ways for each adaptative pattern are presented. If for the island hunters there is a large unknown hiatus, the sea-littoral adaptation process may be traced back instead some 6000 years. Cultural and stratigraphic sequences are presented .

A tantalizing recurrent fact is the apparently infrequent use of littoral resources before the Middle Holocene and the sudden out-break of littoral adaptations in widely separated regions of the World. It has been argued that the sea rise would have swept all the older than Middle Holocene coastal sites, thus the littoral adaptations could have had more ancient roots. The maximum known ages for this tradition in the Beagle Channel coincide with the sea-level stabilization, but the reasons why the reasons of such alternative is considered as improbable as well as the reasons to evaluate the available radiocarbon dates as close to the maximum possible ones are herein discussed.

RESUMEN

En Tierra del Fuego coexistieron dos patrones adaptativos contrastantes: cazadores y recolectores pedestres en la Isla Grande y canoeros adaptados a la vida litoral marítima en los canales y costas isleñas. Se presentan las antigüedades

máximas conocidas de cada patrón adaptativo y sus posibles
vías de poblamiento. Si bien en la secuencia ocupacional de
los cazadores pedestres hay un largo hiatus aún desconocido,
el proceso adaptativo a la vida litoral marítima puede
rastrearse por más de 6000 años. Se presentan las secuencias
cultural y estratigráfica de la región del Canal Beagle.

Un llamativo hecho recurrente es el aparentemente infrecuen-
te uso de recursos litorales con prioridad al Holoceno Medio
seguido de un repentino surgimiento de adaptaciones litorales
marítimas en regiones muy distantes entre sí. Se argumentó que
la elevación del nivel del mar previa al Holoceno Medio habría
barrido todos los sitios costeros pudiendo, por lo tanto, tener
tales adaptaciones raíces más profundas. Las antigüedades máximas conocidas en el Canal Beagle para esta tradición coinciden
con la estabilización del nivel del mar pero se discuten las
razones para considerar dicha alternativa como improbable.
Al mismo tiempo se argumentan razones para considerar a dichas
antigüedades como cercanas a las máximas posibles.

INTRODUCTION

The existence in Tierra del Fuego of two constrasting landscapes
is well-known. Throughout north and northeastern Isla Grande
undulating low plains are present, characteristic of a deeply-
eroded fluvioglacial relief (Codignotto and Malumian, 1981),
with open low vegetation and abundance of peat bogs, streams
and marshes. The eastern and southern parts of the island are
more rugged: there are mountains with crests covered by snow
all the year long, irregular, intended shores and abundant
fiord-like bays, channels and islands. The subantarctic forest
thrives on the slopes, the genus **Nothofagus** being the dominant
element.

These regions were inhabited in the XIXth century by likewise
contrasting (both in physical and cultural aspects) groups
of aboriginals. The northeastern slope of the mountains and
the northern plains were the land of the Selk'nam, whose most
important subsistence resource was the guanaco (**Lama guanicoe**),
hunted with bows and arrows. Most of their food, all of their
clothing and important parts of their dwellings came from this
ruminant. Marine resources were used only as a complement and
these people did not have canoes (Gusinde 1931, 1937; Bridges
1949; Chapman 1986).

134

On the other hand, on the southern and western shores and
surrounding areas lived other people who centered their life
around the canoe; of these, we can distinguish the Yamana
(Beagle Channel to Cape Horn) and the so-called Alacaluf
(around the western mouth of the Strait of Magellan). They
hunted guanaco when possible, but their main subsistence came
from pinniped (**Arctocephalus australis** and **Otaria flavesans**)
hunting and shellfish gathering; bows and arrows were known,
but the favourite weapon was the harpoon. A number of
differences from the inland people could be mentioned one that
strongly attracted the attention of Europeans was that canoe
people could live almost naked in such a rigorous climate
(i.e Darwin 1839). Physically, the Selk'nam were very tall
(average height of males, 175 cm), very robust and of harmonic
proportions; the southern people were much smaller (17 cm less
in average height) but very sturdy (Fitz Roy 1839; Weddell
1825) and their body proportions were disharmonic (to European
eyes). The main sources of data on the Yamana are Hyades and
Deniker (1891) Gusinde (1937), Lothrop (1928), Bridges (1949)
and others.

It is probable that these differences - at least the contrast
between the inland hunters and the canoemen - have existed for
a long time. In the north of the Isla Grande of Tierra del
Fuego there is a proof of human presence as far back as the
Late Pleistocene. At the Tres Arroyos site in Chile (for site
locations, see figure 1), Massone (1983) obtained radiocarbon
dates (on collagen analysis, unfortunately without specifying
whether on terrestrial or marine animals) that place human
occupation by the eleventh millenia BP (Dic 2733: 10420 BP
$\pm$100; Dic 2732: 10280 BP $\pm$ 110). From the lower layers of
the Marazzi rock shelter, near Bahía Inútil, Laming-
Emperaire (1968:67) dated a piece of charcoal at 9590 BP
$\pm$210. Unfortunately, the artifacts recovered in association
with the materials dated were few and of little diagnostic
value, so it is not possible to learn a great deal about the
way of life or filiation of their owners. Although there are
no confirmed data on when the Strait of Magellan opened
completely, it is very probable that this people arrived on
foot, crossing on still unbroken morenic arc.

Thereafter, unfortunately, the sequences of occupation is
unknown; up to the present day, the advances in archaeological
investigation have not been able to close a large hiatus. The
next reliable date appears in the Cabeza de León layer B
dated by Saxon as 1100 AP $\pm$ 95 (MC 1069, cit. in Radiocarbon

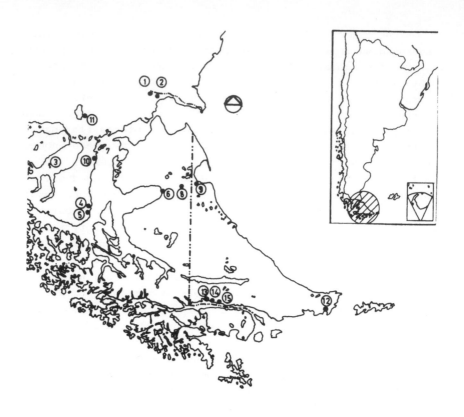

REFERENCES

1 - Fell.
2 - Palli Aike.
3 - Englefield.
4 - Bahia Buena.
5 - Punta Santa Ana.
6 - Marazzi.
7 - Is. Isabel.

8 - Tres Arroyos.
9 - Cabeza de León 1,2,3 y 4.
10 - Cabo Negro.
11 - Cañadón Leona.
12 - Bahía Valentín.
13 - Lancha Packewala.
14 - Túnel Site.
15 - Remolino.

Figure 1. Archaeological sites location

19, n°1, and by Borrero 1981:264). Given this hiatus and the scarce information on the oldest moments, it is not possible to assure whether the guanaco hunters of recent centuries represent direct descendants of the first inhabitants of Tierra del Fuego or not.

The situation is different in the channels and islands of the Magellan-Fuegian area; there it has already been possible

to establish human presence, organized on a dependence on
marine littoral resources, at least throughout the last 6000
years (Orquera and Piana 1983). In the early excavation of the
Englefield site (Otway Sound, southern Patagonia) dates of
9236 BP $\pm$ 1500 and 8446 BP $\pm$ 1500 were obtained (Emperaire
and Laming 1961:16), associated with materials that pointed
to such adaptation, but these results were questioned because
of their great age (Ortiz Troncoso 1979). Another sample from
the same site (**op cit.**), at 3915 BP is questionable because
the 14C analysis was made on burnt bone and appears to be too
recent (Cf. Polach and Golson 1968:25). Very recently, at the
same altitude above sea level but from a nearby different site,
a date of 5500 BP was obtained by Legoupil (1985-1986:43).

On the Brunswick Peninsula (southern Patagonia), Ortiz
Troncoso excavated two archaeological sites, Bahía Buena and
Punta Santa Ana, the former with reliable dates of 5895 BP
$\pm$ 65 (GrN7614), 5770 BP $\pm$ 110 (GrN 7613) and 5210 BP $\pm$ 110
(Gif 2927) (Ortiz Troncoso 1975: 104 and 1980: 181). A somewhat
older date at the second site (GrN 7612: 6410$\pm$ 70;
has a lower degree of certainty because, in contrast to the
others, it was based on marine shells; the magnitude of the
reservoir effect for the Strait of Magellan, which must be
used to correct it, is still unknown.

In the Beagle Channel region, most of the existing information
comes from the excavation of the Lancha Packewaia, Tunel I,
Shamakush I and Shamakush X sites, planned within the framework
of the Proyecto Arqueológico Canal Beagle and carried out by
the authors since 1975 (in the first stages, in collaboration
with Arturo E. Sala and Alicia H. Tapia). During this research
a large quantity of widely varied information on technology,
subsistence, use of space and other aspects of the prehistoric
human adaptation to the environment were accumulated. Part of
this has been presented (Orquera and Piana 1983; Orquera, Piana
and Tapia 1984; Piana 1984; Piana and Orquera 1985 and others)
but a large part is still being processed. In this
investigation we made abundant use of biological and geological
sources and have obtained information that may be of interest
to specialists of these and other sciences.

STRATIGRAPHY AND ANALYSIS

The Túnel site is located on the north coast of the Beagle
Channel at 54°49'15"S and 68°09'20"W. Archaeologically

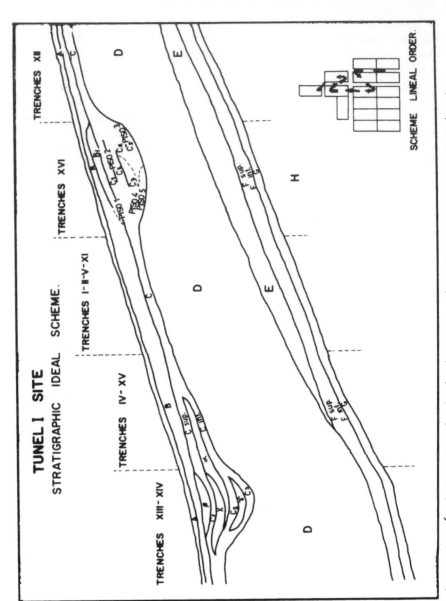

Figure 2. Túnel I Archaeological Site. Stratigraphy sequence present in the different trenches.

fertile layers, reaching a maximum of 180 cm deep, covered a small spur of the mountain slope at 14m/a.s.l.; a steep talus separates it from a terrace level at 5 m/a.s.l. Study of the site started in 1976 and 150 m were excavated during eight field seasons, using completely microstratigraphic techniques.

Even though the research was undertaken with a regional approach - so that it should be based on a series of complementary sites - and though the interest in key sites that supposedly would synthesize the totality of a peopling process is out of fashion, from an archaeological perspective Túnel I must be considered as an important site due to:

1. the quantity and quality of the information yielded;

2. the length of the occupational sequence (a period of human presence near the Beagle Channel four times longer than supposed prior to 1975; cf. Bird 1938:263 and Menghin 1972:30);

3. the changes through time of the sorts of activities carried out at the site;

4. the chronological and functional complementarity with Lancha Packewaia;

5. the refinement of microstratigraphic techniques for shell midden excavation developed at the site, that resulted in a qualitative advance in the data base collected.

The site stratigraphy cannot be reduced to a single scheme because there are significant differences between the northern and southern portions of the site (Figure 2). We may consider the following as a basic scheme (see ^{14}C dates in Table I):

A layer: Present-day humus-like soil, partially mixed with powdered shells. Thickness: up to 2 cm.

B layer: **Nothofagus sp.** bark, product of sawmill activities carried out from ca 1900 to 1960 AD. Thickness: up to 31 cm.

C layer: Humus-like buried soil. Thickness: normally 5 to 7 cm, but in certain places up to 30 cm.

D layer: Shell middens produced by food refuse accumulation (especially **Mytilus** but also other mollusk shells mixed with pinniped, guanaco and avifauna bones; also includes abandoned artifacts and other products of human activity). Thickness: up to 125 cm. This layer is actually composed of a super-position and interdigitation of:

- a large numer of refuse units and shell midden units, each normally a few centimeters thick and ranging in diameter from decimeters to a few meters;

- very thin, discontinuous, not extended lenses of humus-like soil (seldom reaching more than a meter in maximum length);

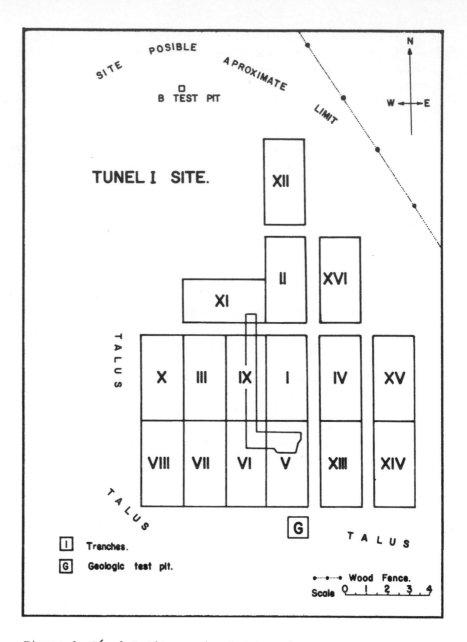

Figure 3. Túnel I Site: excavated trenches.

- human occupational floors, on either the humus lenses or
discordant surfaces between the shell midden units;
- fine lenses consisting of thermoaltered sediments (by fires
lit over the occupied surfaces mentioned above) and by
dispersal of charcoal and ashes.

E layer: black, somewhat unctuous silt, intercalated with
litters of small pebbles of 8 to 32 mm in length;

F layer: sandy and clayish silt, gray to black (when wet)
in the upper part and reddish brown in the lower;

G layer: black silt when wet and reddish brown, darker than
lower F when dry;

H layer: basal surface of the archaeological site.

This layer was not excavated but in a test pit Dr. Jorge
Rabassa (Pers. Comm.) was able to distinguish: H_1: yellowish
aglutinated sand, enclosing small pebbles, with cobbles (5 to
30 cm in diameter) in the upper portion, which emerge from the
surface; probably a result of colluviation of materials from
the lateral moraine of the last glacial advance, which
parallels the Beagle Channel coast some hundred meters north
of the Túnel site. H_2: Loose nonaglutinated sand with or
without dispersed pebbles. H_3: beach gravel (lacustrine?).
H_4: very hard consolidated sand with small pebbles (till?).

Layers B and D are predominantly anthropogenic. The other
layers have a prevailing natural origin, even though C, E and
F include material introduced by human activity. Layers A and
B also include such materials, but in very low quantities and
presumably redeposited. Layers G and H are archaeologically
sterile.

In the southeastern quadrant (trenches IV; XIII; XIV and
XV; see Figure 3) the C layer is divided in two; three and
even four very thin but continuous small layers due to
anthropogenic shell midden lense intercalation. Consequently,
in that sector within B and D layers the following super-
position was found:

Co layer: almost imperceptible lens of grayish soil.

Beta lens: shell midden. Thickness: up to 11 cm.

C_1 layer: clodded grayish humified silt. Thickness: from
1 to 5 cm.

X/ alpha lens: shell midden. Thickness: up to 22 cm.

C_2 layer: similar to C_1. Thickness: 1-5 cm.

Gamma lens: shell midden. Thickness: up to 24 cm.

C_3 layer: similar to C_1 and C_2. Thickness: 1 to 10 cm.

On the other hand, in the northeastern section of the site

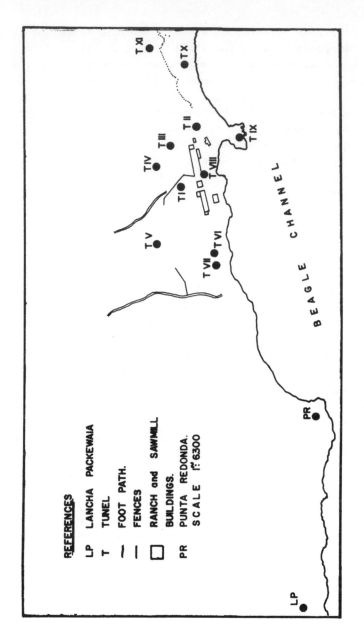

Figure 4. Location of the archaeological sites in Túnel Locality

(trench XVI, see Figure 3) the C layer was also found to be subdivided, not for shell midden intercalation but for:

1. erosion (rain wash and aeolian) and redeposition events.
2. interposition of human occupational floors.

In consequence, in that sector, within the C layer, the following were differentiated:

C_1: humified dark silt.

Piso 2: occupational floor, very subtile powdered shell lens with artifacts.

C_2: similar to C_1.

C_r: colluviated humified silt, includes occupational floor "Piso 3", undisturbed although the surrounding colluviation was due to the very brief developing time of the occupational floor.

C: similar to C_r but with a higher density of powdered shells (probably washed or blown from nearby shell middens). Both C_r and C filled a previous depression partially shaped by preexisting shell middens and later deepened by water erosion.

C_3: dark humified silt, similar to C_1 and C_2; occupational floors 4 and 5 were found included. This layer was sectioned by the previously mentioned erosion furrow (later filled with C_r and C).

Among the layers of mainly non-anthropogenic origin, E layer has a very limited horizontal extension outside the archaeological site limits. Others, such as F, on the contrary, have a much larger expansion and form the ground on which the surrounding **Nothofagus** forest roots (J. Rabassa, pers. comm.).

A sedimentological analysis of a first series of 16 samples from Túnel was conducted by Dr. María Clara Etchichury and Lic. Roberto Gualzzetti (Universidad de Buenos Aires); a second one is under study by Lic. Mónica Falcone and Lic. María Elena Forzinetti. From the first report (Etchichury and Gualzzetti MS), the following can be drawn:

1. the highest humus accumulations are localized in layers A (66%) and F (61%). High values were also found in C_3 (SW quadrant) and G layers. This suggests the existence of at least four superimposed edaphic horizons, which is corroborated by the assiciation with vegetal fibres, spores, pieces of cutin, pollen, phytolithes, etc. It must be mentioned that no analysis was done on C_2 (SW quadrant) nor on the septentrional C units. Other underlying units showing decreasing humic concentrations could indicate percolation processes.

Table 1. Radiocarbonic dates of Túnel I sites

Lab.	Bed	Material	Date	Obs.
Beta 4388	beta	charcoal	450 BP $\pm$ 60	
AC 701		charcoal	670 BP $\pm$ 80	
AC 850	C$_1$ (Floor 2)	charcoal	1920 BP $\pm$ 80	
AC 851		charcoal	1990 BP $\pm$ 110	
AC 852	not determ.C	charcoal	2000 BP $\pm$ 110	
AC 1030	alpha	charcoal	2660 BP $\pm$ 100	
Beta 2516		charcoal	2690 BP $\pm$ 80	
AC 677		charcoal	3030 BP $\pm$ 90	
Beta 4387	gamma	charcoal	2880 BP $\pm$ 60	
AC 702	lower C	charcoal	3530 BP $\pm$ 90	
Beta 4385		charcoal	4300 BP $\pm$ 80	
AC 833	D	charcoal	4590 BP $\pm$ 130	dubious
AC 683		charcoal	5360 BP $\pm$ 120	
AC 236		charcoal	5700 BP $\pm$ 170	
CSIC 308	D	charcoal	5850 BP $\pm$ 70	
AC 838		charcoal	5950 BP $\pm$ 170	
CSIC 309		charcoal	5960 BP $\pm$ 70	
AC 1028		charcoal	6020 BP $\pm$ 120	
Beta 2819		charcoal	6140 BP $\pm$ 130	
AC 840		charcoal	6410 BP $\pm$ 150	
AC 842		charcoal	6750 BP $\pm$ 160	dubious
AC 238		charcoal	5690 BP $\pm$ 180	
CSIC 310	E	charcoal	6070 BP $\pm$ 70	
Beta 3270		charcoal	6200 BP $\pm$ 100	
AC 674	lower F	charcoal	6680 BP $\pm$ 210	
Beta 2517		charcoal	6980 BP $\pm$ 110	
AC 660	G	organic mud	6820 BP $\pm$ 140	
AC 660		organic mud	6840 BP $\pm$ 120	

2. in lower F and G layers there are high proportions of volcanic glass (60% and 40% respectively) with little evidence of transport alteration; the other layers show neap percentages (1% or just traces).

At the nearby Lancha Packewaia site, at a distance of 1 km (54°49'15"S and 68°09'45"W, see Figure 4) and located on a small terrace level at 6-7 m/a.s.l., stratigraphy is as follows:

A layer: present-day dark grey humic soil. Thickness: 1 to 15 cm.

B layer: composed shell midden. Thickness: up to 61 cm.

C layer: gray humus-like soil. Thickness: 1 to 11 cm.

D layer: composed shell middens. Thickness: up to 15.

D layer: gray humus-like soil. Thickness: up to 15 cm.

X layer: compressed shell midden. Thickness: up to 32.

E layer: beach sand and gravel. J. Rabassa reported that its origin is not marine, as previously thought (Orquera 1978) but fluvial, developed by a stream lateral to the site.

Radiocarbon dates (Table I and II) indicate that all this sediment bulk is comparable in age with C layer of the Túnel I site and the intercalated shell midden lenses.

Dr. Calvin Heusser (New York University) undertook the palynological analysis of the samples taken from different layers of Túnel I and Lancha Packewaia sites (Table III). Analyzable pollen was not preserved in the shell middens nor in the carbonaceous samples such as of C_3 of Túnel I. Since these were sites with intense human occupation, pollen results may not sharply reflect regional environmental conditions. Nevertheless, the path from an initial absence of arboreal vegetation to an overwhelming dominance of **Nothofagus sp.** species (from 6000 BP to present) is shown as a noticeable tendency. Two circumstances to mention in relation to the palynological record are:

1. the scarcity of pollen of Gramineae (4% and 3%) in the samples from F and G, even though the sedimentological analysis pointed to the presence in those layers of relatively high quantities of cuticles and phytolites;

2. the lack of **Berberis sp.** pollen in the palynological record, even though this genus is relatively well represented in the surroundings at present.

Preliminary analyses of the $^{16}O/^{18}O$ content in marine mollusk shells carried out by Lic. Héctor Panarello (INGEIS, Buenos Aires) suggest that the average water temperature in

Table 2. Radiocarbonic dates of Lancha Packewaia site.

Lab.	Bed	Material	Date	Obs.
M 1062	B	charcoal	280 BP ± 85	
M 1063		charcoal	280 BP ± 85	
MC 1066	C	charcoal	410 BP ± 75	
NC 1063	C	charcoal	455 BP ± 85	
CSIC 314		charcoal	470 BP ± 50	
MC 1065	D	charcoal	1080 BP ± 100	
CSIC 311		charcoal	1120 BP ± 50	
CSIC 312		charcoal	1590 BP ± 50	
CSIC 306	X	charcoal	4020 BP ± 70	
CSIC 305		charcoal	5920 BP ± 90	discarded
MC 1068		charcoal	4215 BP ± 305	minimum age
CSIC 307	E	bone	4980 BP ± 70	

the Beagle Channel could have been ± 1°C lower some 6000 years ago and ± 1°C higher some 2800 years ago than at present (Panarello 1986). Although these values were obtained as an average of the mollusk's life span (in this case, at least an estimated 3-5 years) and comparison with present temperature was through the $^{16}O/^{18}O$ contents of present-day shells, it must be pointed out that the differences found are within the range of variation of water temperatures in the Beagle Channel registered for the 1984-1985 and 1985-1986 yearly cycles (C. Schroeder, pers. comm.).

Other analyses on archaeological shells from Túnel I and other coastal sites along the Beagle Channel (Albero **et al.** 1985-1986) show:

1. that the reservoir effect in marine mollusk and mammals dated by radiocarbon analysis has been estimated at ± 560 years for the last 6000 years;

2. the great variability of response to ^{14}C analysis of

different marine shells, with the lowest reliability in
Balanus and **Trophon.**

CULTURAL SEQUENCE

The characteristics of the archaeological remains found in the
sites excavated by us and their age differences enable us to
differentiate the following cultural components.

1 FIRST COMPONENT OF TUNEL (LOWER F LAYER)

This component corresponds to a short occupation some 7000
years ago (Beta 2517) at some time during the autum
(determination by Lic. Enrique Crespo (CENPAT) on thin cuts
of cannine teeth of **Arctocephalus australis;** the pinniped was
consume during the short occupation). It should be
remembered that at that time the land around Túnel was not
yet covered with forests, but with herbaceous plants and
cryptogams. Selection of the place for human occupation
probably was determined by the presence of a small 80 cm ridge
that offered protection for their fire from the southwest -
nowadays predominant - winds. The implements of these people
which were conserved reveal great technical skill (Orquera
et al. 1982, 1984; Piana 1984) although their tantalizing
characteristics do not permit precise determination of
cultural filiations, it is possible to assert:
   1.1 they were apt for hunting and butchering terrestrial but
not marine mammals. Nothing within the tool kit implies so
specialization or even skill in this latter class of
subsistence .
   1.2. that, nevertheless no guanaco remains appeared at the
site - although the open environment then present should have
been favourable for the occurrence of guanaco; the archaeo-
faunal record points to an incidental usufruct of marine
littoral resources.

2 SECOND COMPONENT OF TUNEL (UPPER F, E AND D LAYERS)

These layers represent a large series of short discontinuous
occupations before and around 6000 BP when the site was

Table 3. Palynological Analises of Túnel I and Lancha Packewaia Samples.

| | Túnel I Site | | | | | Lancha Packewaia Site | | |
	G bed	lower F bed	E bed	C₂ bed	C₁ bed	D' bed	C bed	A bed
Seed plants								
Nothofagus	11	26	97	100	100	91	69	81
Empetrum	15	8	–	–	–	–	–	–
Embothrium	–	–	–	–	–	–	–	+
Gramineae	3	4	1	–	–	+	+	+
Caryophyllaceae	–	–	1	–	–	–	+	3
Gunnera	–	–	1	–	–	–	–	–
Ligaliflorae	2	+	–	–	–	–	+	2
Tubuliflorae	68	60	–	–	–	8	28	11
Cryptogams								
Polypodiaceae	87	31	4	+	+	7	3	+
Lycopodium	9	18	–	–	–	+	–	–

% for seed plants excludes cryptogams spores

% for cryptogams is based on total count of seed plants + cryptogams

+ is less than 2%

already covered by dense **Nothofagus** forest (Table III). In contrast to the former the new settlers were already markedly specialized in their dependence on the marine littoral resources, mainly in the capture of pinnipeds and - although not at the earliest stages - sessile marine invertebrates. Studies carried out by Lic. Adrian Schiavini (CADIC-CONICET, Ushuaia) on pinniped remains indicate that the site was occupied during different seasons and that the archaeological record shows a notable predominance of fur seals (**Arctocephalus australis**, 95%) over southern sea lions (**Otaria flavescens**, 5%). Hunting of guanacos, cormorants (**Phalacrocrorax** sp.) and penguins contributed complementary resources to the diet. By that time, normal domestic activities that can be expected at a base camp were carried out at Túnel I. Ample use was made of grey and greemish silicified slates of acid volcanic (probably rhyolitic) origin for chipped stone implements (M. Falcone, pers. comm.). Much less use was made of claystone, basalt and other rocks; amphibolites were preferred for objects smoothed by pecking.

3 ANCIENT COMPONENT OF LANCHA PACKEWAIA (E AND X LAYERS)

Although at another site, dated $\pm$ 4300-4000 BP, the life style and encampment type of the previous component were maintained with only a few variations. The most notable difference was in the increment in guanaco capture (reaching $\equiv$ 33% of the calories consumed, Saxon 1979), coinciding with the first appearance of flaked stone spear heads. Andesitic basaltic vulcanite, a rock which is alochtonous to the site, was used for these arms (J. Morellim pers. comm.). The rest of the stone implements were shaped in raw materials similar to the ones described for the Second Component.

4 THIRD TO SIXTH COMPONENTS OF TUNEL (C , X/ALFA, PISO 2 AND BETA LAYERS)

These components represent a series of short occupational events assignable to the same tradition of littoral adaptation. Activities registered in the archaeological record point to specialized encampments oriented toward guanaco hunting and butchering as a complementary activity to the constant predation on pinnipeds and mollusks.

5 RECENT PHASE OF THE BEAGLE CHANNEL (B, C AND D LAYERS OF LANCHA PACKEWAIA, SHAMAKUSH I AND X AND THE ABOVE MENTIONED SIXTH COMPONENT OF TUNEL)

The littoral adaptive form was conserved although there were changes in the morphology and technology of certain types of artifacts. As far as it is presently known, bows and arrows were introduced during this time. At Lancha Packewaia in the XVIIth century AD, pinnipeds made up 91% of the calories consumed; barely 4% was obtained from guanaco meat. Among the former, the preference for **Arctocephalus australis** continued (94% of MNI of 99% of the total of bones identifies, Saxon 1979). It should be noted that this species is rarely seen in the eastern Beagle Channel at present.

DISCUSSION AND CONCLUSIONS

In our excavations, in consequence, we have found evidence of two different human adaptative systems: one of terrestrial hunters (First Component of Túnel) and the other of littoral hunters and gatherers (all of the other archaeological manifestations described). According to the accumulated knowledge on the prehistory of this region to date, the inland hunters did not become successful in settling the coastal environment, while remains of its occupation by the second system are recurrently found throughout the last six thousand years. This latter adaptation was similar to the one produced in the western Magellan region and adjacent environments and it seems justifiable to include them both in an adaptative Cultural Tradition of the Magellan-Fuegian Islands and Channels (Orquera **et al.** 1978, 1983). This tradition had coincidence points with several generically similar adaptative processes in other regions of the world that counted on the availability of similar subsistence resources (i.e. British Columbia, the Aleutians, parts of New England, etc.).

During recent years, there has been an increasing current of archaeological interest in the coastal orientation of some groups of hunters and gatherers (i.e., Osborn 1977; Yesner 1980, 1983; Perlman 1980; Moseley and Richardson 1983 among others). This orientation, in effect, permits fruitful comparisons and insights in respect to the strategies of human survival. In the course of such studies, a tantalizing

recurrent fact is the apparently infrequent use of littoral
resources prior to the Middle Holocene and then the sudden
outbreak of littoral adaptations that occurred almost
simultaneously in widely separated regions of the world.
Among the several possible interpretations, one sustains that
the coastal encampments established prior to the Middle
Holocene were swept by the rising level of the sea, which
began at the end of the Pleistocene and stabilized about
6000-5500 BP (Fairbridge 1962; Morner 1971; Milliman and
Emery 1968 among others). The conservation of the more recent
coastal sites, due to the stabilization of the sea level,
would have given rise to the belief in an apparently sudden
bloom while, actually, the adaptative process to a littoral
environment could have had deeper roots. This interpretation
was sustained by Hardy (1960), Sauer (1962) (both cited by
Yesner 1980:733), Perlman (1980;285), Rogers (1980:742),
Moseley and Richardson (1983:2), and others. In opposition,
Yesner (1983:1-2), Custer and Steward (1983:2) and
- implicitly - Osborn (1977) think that human littoral
adaptation is a late phenomenon, and so that the sea rise
would have not acted upon the preservation of related
archaeological sites.

   In the Beagle Channel the oldest evidence of coastal
adaptation known to date (Second Component of Túnel, slightly
earlier than 6000 BP) coincides with the beginning of such
sea level stabilization, so one must question if it could be
possible that in this region there had been even earlier
human groups assignable to the adaptative tendency which we
have been discussing, but whose encampments are now impossible
to find. None the less, there are motives for considering this
improbable. The reasons are as follows:

1. Site survival against eustatic rise

It has long been known that in this region the land is rising
in relation to the level of the sea. Urien (1966) presented
data from which it was possible to deduce an average rate of
ascent during the last 5500 years of approximately one meter
each 600 years. Other data, obtained by Porter **et al.** (1984),
would suggest a much slower rhythm, but recent studies by
Rabassa, Heusser and Stuckenrath (1986) hint once again that
higher values (a meter each 500 years, at least for the last
4400 years) could be accepted. It seems likely to imagine that,
whether with the same rate or another one, the uplifting would

151

had been producing for a considerable lapse even previous to that time. If this is so, the possibility that coastal encampments were covered by eustatic rise which ended in the Middle Holocene is greatly but not totally diminished.

To calculate such eustatic movement, we can use the curves proposed by Fairbridge (1962) and Morner (1971). It is true that the former has been questioned in regard to the supposed existence of levels higher than at present at different moments after 6000 BP (cf. Morner, 1971; Thoms **et al.**, 1969; Shepard **et al.**, 1967). In addition, if we apply the Fairbridge curve to the uplift of the north coast of the Beagle Channel, the Lancha Packewaia site would have been covered by the sea soon after its first occupation (4300-4000 BP), an event which never took place. Therefore, we will not take into account the supposedly high sea levels of the sixth and later millenia BP with regard to the period previous to 6000 BP, both curves are sensitively coincident and does not appear to be major disagreements.

During the period from 7000 to 6000 BP, the rise in sea level would have been six meters. The resulting rhythm is noticeably higher than the possible land rise, so that a coastal site would have become rapidly submerged or destroyed by wave action. Let us suppose for this period a rate for land uplift of - in interest of precaution - only half of the one suggested by Rabassa **et al** (1986): in other words, a rhythm of a meter each thousand years. If both movements were matched, a hypothetical site in 7000 BP located at only 5 m/a.s.l. would have been saved from the posterior rise in sea level. If this site had been occupied in 8000 BP or in 6500 BP, the requirement would have been somewhat higher (in the second case, because of the slight inflection shown by the Fairbridge curve after 7000 BP): in order to be saved from being washed out by posterior wave action, the hypothetical 8000 BP site would have gad to have been at 12 m/a.s.l. at that time, and that of 6500 BP at 7.50 m.

Such situations would not have been very improbable, especially after 7000 BP:

1.1 It is true that the majority of sites of littoral hunters and gatherers tend to be - for obvious reasons of convenience - near the coast. But on the Beagle Channel, at least, it is unlikely that settlements were located below 3 m/a.s.l.: dwelling huts must be safe from extraordinary tides and from the surge during days of storm; they also must have a certain perspective to descry at a distance, and take advantage of soft grass rather than beach pebbles. Therefore, the strip of potential occupation subject to the risk of posterior destruction is reduced.

1.2 In the Beagle Channel region, the coast is commonly steep or with cliffs. This leads to three effects: a) zones with gradients appropriate for human settlements are few and are now easily recognizable; b) the rise in sea level could not be translated into considerable horizontal transgression (so that the location of the coastline would not undergo great variation in this direction); and c) when there are no zones of low altitude appropriate for human occupation, the people were forced to install their dwellings high above the declivity.

Therefore, it is erroneous to consider that all settlements of littoral-adapted hunters and gatherers must be on sea-shores at a low altitude above sea level. In fact, there are many cases to the contrary:

a) the final occupations of Lancha Packewaia were produced when the site was already at 6 to 7 m/a.s.l.

b) the later occupations of Túnel I occurred when the site was at 15 m/a.s.l.; at the same time, Túnel II (at 20 m/a.s.l., 1230 BP $\pm$ 110) and Túnel III (at 25 m/a.s.l., from 1130 BP $\pm$ 120 to 420 BP $\pm$ 80) were also occupied while there were suitable places at a lower altitude.

Other non-dated but presumably recent sites, are located at 30 or more meters above present sea level (i.e. Túnel XI). In consequence, the assertion that the eustatic rese previous to 6000 BP had washed out all the remains of earlier hypothetical settlements of littoral hunters and gatherers has very little credibility.

2. The forest expansion

Nevertheless, it is possible to assume that these high older sites could exist and that our ignorance of them is only due to insufficient surveys. Notwithstanding, the settlement by littoral adapted people requires that several previous conditions be satisfied:

2.1 existence of abundant littoral biomass on which predation can be made for subsistence;

2.2 climatic conditions must not be too extreme. Even though
the external islands, razed by hurricane winds and rainfall
measured in thousands of millimeters, were occasionally visited
in search of special resources, the fundamental part of the
life of the canoe aboriginals which inhabited the region under
study was spent in sounds interior channels, and other less
turbulent waters and with more moderate precipitation (Laming-
Emperaire 1966:302-303).

2.3 existence of available extended woods of large trees.

The first condition required, in turn, the presence of an
opening to sea. Unless anadromous fish are present - which
is not the case in Tierra del Fuego - it is almost impossible
for a lacustrine basin to supply enough food so that the
coast-benefits relation can permit the development of human
 reliance on such environment. On the contrary, cold climate
maritime littorals are particularly rich, especially if there
are channels or narrows between islands; the upwelling of cold
waters bringing nutrients to the photic zone allow complex
trophic chains (Sander and Steven 1973, cit. Yesner 1980).
    The fact that the hunters who passed through Túnel some
7000 years ago consumed one **Arctocephalus** (see above) implies
that the Beagle Channel was open to the sea, at least in one
of its extremes. Rabassa **et al.** (1986) were able to confirm
that in 9380 BP $\pm$ 105 today's Beagle Channel was perhaps still
partly occupied by a lake and estimated that its opening
occurred before 8240 BP $\pm$ 60. Since this date was obtained
from sea shells, it is possible that such age should be reduced
in some - still undetermined - proportion due to the Reservoir
Effect (cf, Albero **et al.** 1985, 1986).
    In consequence, the sea and littoral resources were already
in the Beagle Channel a thousand or more years before the sea
stabilized at the present level and before the oldest presence
yet known of the littoral adapted hunters and gatherers (Second
Component of Túnel).
    In contrast, the third condition (requirement of the existence
of forests) seems to be more decisively opposed to the
possibility of finding much older traces of this sort of
adaptation of these people on the shores of the Beagle Channel.
The technological requirements of counting on a reliable, well-
dispersed, and readily available source of well-grown trees for
canoe making and long straight stems for harpoon shaft has been
pointed out (Orquera **et al** 1984; Orquera and Piana in press:

Piana 1984). A similar nexus, but for other reasons, was proposed for British Columbia by Hebda and Mathews (1984). Actually, such requirements cannot be satisfied by bushes or scrubby trees with thin trunks or lacking stems of adequate length; also, to bring appropriate wood and bark from great distances would result in prohibitive cost.

Such requirements do not greatly affect the inland hunters who inhabited the northeastern steppe regions of Tierra del Fuego or the Patagonian plateau. They had no need of canoes; the appropriate weapon for guanaco hunting was bow and arrows whose raw material may be taken from ñires (**Nothofagus antarctica**) and other scrubby trees or bushes (cf. Chapman 1986:58 and other authors mentioned there); and only housing imposed the possession of wooden poles (a necessity that was covered by reducing housing to transportable windbreakers).

The strength of the forest conditioning on the human littoral adaptation is supported by the fact that, at the Strait of Magellan, Emperaire (1955:78) affixed the canoers' expansion limits at Cabo Negro, which corresponds to present forest extension. It is true that there are archaeological remains of littoral-adapted people at Isla Isabel, a short distance to the east, in spite of the fact that there are no trees there (Bird 1980); but the distance from Cabo Negro is short and remains there could pertain to excursions in search of penguins. In the south of Tierra del Fuego Martial (1888:184) and Cooper (1946:181) stated that the Yamana land extended to the eastern mouth of the Beagle Channel; Spegazzini (1882), more specifically, pointed to Bahía Sloggett, while Bridges (1894: 61) and Lothrop (1928:116) extended such territory to Bahía Aguirre. Nevertheless, Martial attributed the land east of the Cabo San Pío to the pedestrian hunters and Bridges himself (1949:34) remarked that the canoers seldom reached Puerto Español. On Península Mitre, the coastal fringe is also wooded, but - in contrast to the channels and islands - is totally unprotected against the Antarctic winds and the rough waters make the use of canoes problematical.

Now then, it has been pointed out that at the Túnel site the **Nothofagus** forest became dominant during the course of the seventh millenium BC; in 7000 BP open countryside and non-arboreal vegetation still existed. This datum is consistent with the very interesting ones obtained by Heusser (1984) from the Puerto Williams profile: until 10080 BP $\pm$ 140 there does not appear to have been any trees nearby and the installation of the **Nothofagus** would have taken place sometime between this date and 5520 BP $\pm$ 70. It is true that in the published graph (Heusser 1984), at first glance - since the vertical column is arranged by depth - the event seems to be contiguous with the former date. But to accept this is to accept an even speed for the sedimentary process, a fact that the same graph shows to be erroneous. Also, to accept that the **Nothofagus** dominance could have been reached at - 10000 BP raises another problem: Heusser's surface studies in Patagonia showed that (Heusser 1984); Túnel is only at 47 km from Puerto Williams and such pollen incidence in layers G and F may be seen in Table III. Instead, if an uneven sedimentation speed - as the graph mentioned above shows - or a discontinuity in the sedimentation process (and therefore the existence of an up to now undetermined temporal hiatus) are to be suspected, all of the palynological reports match.

Rabassa **et al** (1986) presented data that suggest a somewhat earlier forestal approximation to the Beagle region: a profile at Isla Gable shows 31% of **Nothofagus** pollen in 9380 BP $\pm$ 105 (SI 6732) and another from Lapataia shows 42% for 8240 BP $\pm$ 60 and 62% for 7260 BP $\pm$ 70 (SI 6737 and 6738). Nevertheless, this discordance may be more apparent than real. The distance between Lapataia and Túnel, 35 kms, does not seem to be a sufficient explanation because of the previously mentioned wind transport range of such pollen. From our point of view, it is more probable that the cause should be looked for in the degree of certainty of the radiocarbon dates envolved: those from Lapataia were obtained on **Mytilus** shells, necessarily affected by Reservoir Effect (Albero **et al** 1985). There is no way to precisely calculate the incidence of the Reservoir Effect in the Beagle Channel for the ages dealt with herein. Notwithstanding, if it is accepted that some degree of incidence must exist (cf. the values known for 6000 BP) the age of the samples 4 and 3 of Lapataia should be younger, thus mearing or concurring with the forestal expansion range that according to the Túnel dates would have happened after 7000 BP. This correction does not affect the sample from Isla Gable because it is dated on wood, but the 31% of **Nothofagus** pollen does not seem to necessarily imply such a well established dense forest as the one required.

If a successful human adaptation to the marine littoral of the Magellan-Fuegian channels and island region requires an extensive, well-settled forest with several decades old trees it may be inferred:

a) unless palynological studies demonstrate that woods with

such characteristics were established quite some time earlier,
it is not probable that the colonization of the Beagle Channel
by the littoral adapted people started much earlier than the
Second Component of Túnel (it is to be remembered that the
people of the First Component of Túnel were inland hunters
and gatherers).

b) on the contrary, it is feasible that human adaptation to
the marine littoral in the region started earlier in the
southwesternmost part of Patagonia and the western mouth of
the Strait of Magellan because the requirements discussed above
were present and the postglacial reimplantation of the forest
would have been earlier. According to the data obtained by
Heusser (1984:67) in the Puerto Hambre profile, between
10940 BP $\pm$ 70 and 7980 BP $\pm$ 50, a gradual increase in the
percent point of domination from an imprecise, but older than
the Beagle Channel date.

This supposition needs checking, something that has not yet
occurred. Evidence found at Túnel I of some sort of contact
between the two regions is too late to suggest a displacement
direction - if there was any. Nevertheless, the proposition
that we would have reached a very proximal moment to the first
settlement on the region of the Beagle Channel of human groups
thus adapted, appears to be well enough sustained to derive
another proposition: the already accumulated archaeological
record from the sites excavated by the Proyecto Arqueológico
Canal Beagle and the ethnohistorical data available would
cover almost the total length of a successful adaptation
process.

Nevertheless, the way of living of way human group cannot
be reflected in all its variety and flexibility in only one or
two sites, no matter how rich the information they may yield.
Our future investigation steps will be directed to more precise
knowledge of the profit and the complementation aspects. These
groups took from the environment between habitats with
different resources.

ACKNOWLEDGEMENTS

We thank J. Rabassa for the geological data on Túnel locality
and R.N.P. Goodall for the critical reading of the manuscript.
We are also grateful to C. Heusser for the palynological
analysis of Lancha Packewaia and Túnel I samples; to M.C.
Etchichury, R. Cualasetti, M. Falcone and M.E. Forzinetti

for the sedimentological analyses; to E. Crespo and A.
Schiavini for seasonality data; to INGEIS for   C dates; to
the Servicio de Cartografía del CADIC for the illustrations;
to the Asociación de Investigaciones Antropológicas and
CADIC-CONICET for their support; and to the many students
and colleagues who have worked or are working with the Proyecto Arqueológico Canal Beagle.

REFERENCES

Albero, M.C.; Angiolini, F.E. & Piana, E.L. 1985. Discordant
  ages related to reservoir effect of associated archaeological
  remains from Túnel site (Beagle Channel, Argentine Republic).
  XII International Radiocarbon Conference (Trondheim, Noruega)
Albero, M.C., Piana, E.L. & Angiolini, Fernandi E. 1976.
  Holocene ^{14}C reservoir effect at Beagle Channel (Tierra del
  Fuego, Argentina). **Quaternary of South America and
  Antarctic Península.** Vol. IV.
Auer, V. 1959. The Pleistocene of Fuego-Patagonia. Part II:
  the history of the flora and vegetation. **Annales Academiae
  Scientiarum Fennicae,** serie A III (Geologica-Geographica)
  50, Helsinki.
Bird, J.B. 1938. Antiquity and migrations of the early
  inhabitants of Patagonia. **Geographical Review.** XXVIII:258-
  275.
Bird, J.B. 1980. Investigaciones arqueológicas en la isla
  Isabel (Estrecho de Magallanes). **Anales del Instituto de
  la Patagonia,** 11:75-87. Punta Arenas, Chile.
Borrero, L.A. 1981. Excavaciones en el alero Cabeza de León
  (Isla Grande de Tierra del Fuego). **Relaciones de la Socie-
  dad Argentina de Antropología,** XIII(1979):255-271. Buenos
  Aires.
Bridges, E.L. 1949. **"Uttermost part of the earth".** New York.
Chapman, A. 1986. **"Los Selk'nam (la vida de los onas)".**
  Emecé Editores, Buenos Aires.
Codignotto, J.O. & Malumian, N. 1981. Geología de la región al
  norte del paralelo 54°S de la Isla Grande de Tierra del
  Fuego. **Revista de la Asociación Geológica Argentina** XXXVI(1):
  44-88.
Custer, J.F. & Steward, R.M. 1983. Maritime adaptation in the
  Middle Atlantic Region of eastern North America. **48th annual
  Meeting of the Society for American Archaeology.** New World
  Maritime Adaptations Symposium (Pittsburgh).

Darwin, Ch. 1839. Journal and Remarks (1832-1836). In: **Narrative of the surveying voyages of his Majesty's ships Adventure and Beagle between the years 1826 and 1836...,** III. Henry Colburn, Londres.

Emperaire, J. 1955. **"Les nomades de la mer".** Gallimard, Paris.

Emperaire, J. & Laming, A. 1961. Les gisements des iles Englefield et Vivian dans la mer d'Otway (Patagonia australe). **Journal de la Société des Américanistes,** 50:7-77. Paris.

Etchichury, M.C. & Gualzetti, R. -M.S. - Sedimentología de muestras de un depósito del sitio Túnel. Territorio Nacional de Tierra del Fuego. 1984. In press, CADIC.

Fairbridge, R.W. 1962. Eustatic changes in sea level. In: **Physics and Chemistry of the Earth.** L.H. Ahrens et al.(eds.), 4:99-185.

Fitz Roy, R. 1839. Proceedings of the second expedition (1831-1836) under the command of captain Robert Fitz Roy, R.N. In: **Narrative of the surveying voyages of His Majesty's ships Adventure and Beagle between the years 1826 and 1836...** II, Herry Colburn, Londres.

Gusinde, M. 1931. **"Die Feuerland Indianer".** I: **"Die Selk'nam".** Modling-Wien.

Gusinde, M. 1937. **"Die Feuerland Indianer".** II: **"Die Yamana".** Modling-Wien.

Hardy, A. 1960. Was man more aquatic in the past?. **New Scientist** 17:642-45.

Hebda, R.J. & Mathewes, R.W. 1984. Holocene History of Cedar and Native Indian Cultures of the North American Pacific coast. **Science** 225(4663):711-713.

Heusser, C.J. 1984. Late Quaternary Climates of Chile. In: **Late Cenozoic Palaeoclimates of the Southern Hemisphere,** J.C. Vogel (ed.), 59-83. Rotterdam, A.A. Balkema.

Hyades, J.D.J. & Deniker, J. 1891. Antropologie et Etnographie. In: **Mission Scientifique du Cap Horn (1882-1883),** VII. Paris.

Laming-Emperaire, A. 1966. Quelques étapes de l'occupation humaine dans l'extreme sud de l'Amérique australe. **Actas y Memorias del XXXVII Congreso Internacional de Americanistas (Mar del Plata, Argentina),** III:301-313. Buenos Aires.

Laming-Emperaire, A. 1968. Missions archéologiques francaises au Chili austral et au Brésil méridional: datations de quelques sites par le radiocarbone. **Journal de la Societe des Americanistes,** 57:77-99. Paris.

Legoupil, Dominique. 1985-1986. "Los indios de los archipiélagos de la Patagonia: un caso de adaptación a un ambiente adverso. **Anales del Instituto de la Patagonia**, vol. 16:45-52.

Lothrop., S.M. 1928. **The indians of Tierra del Fuego.**
Museum of the American Indian, Heye Foundation, New York.

Martial, L.F. 1888. Histoire du voyage. In: **Mission Scientifique du Cap Horn (1882-1883). I.** Paris.

Massone M., M. 1983. 10.400 años de colonización humana yámana en Tierra del Fuego. **Infórmese**, 14.

McGhee, R. 1983. Prehistoric Maritime Adaptations in Artic and Subartic Canada. **48th Annual Meeting of the Society for American Archaeology. New World Maritime Adaptations Symposium** (Pittsburgh).

Menghin, O.F.A. 1972. Urgeschichte der Kanuindianer des südlischsten Amerika. **Anales de Arqueología y Etnología** XXVI: 5-51. Mendoza, Argentina.

Milliman, J.D. & Emery, K.O. 1968. Sea levels during the past 35.000 years. **Science**, 162:1121-1123.

Morner, N.A. 1971. Eustatic changes during the last 20.000 years and a method of separating the isostatic and eustatic factors in an uplifted area. **Palaeogeography, Palaeoclimatology, Palaeoecology**, 9(3):153-181.

Moseley, M. & Richardson, J.B. III. 1983. Fishing versus farming in the Peruvian preceramic. **48th Annual Meeting of the Society for American Archaeology, New World Maritime Adaptations Symposium** (Pittsburgh).

Orquera, L.A. & Piana, E.L. 1983. Prehistoric maritime adaptation at the Magellan-Fuegian littoral. **48th Annual Meetting of the Society for American Archaeology, New World Maritime Adaptations Symposium** (Pittsburgh).

Orquera, L.A. & Piana, E.L. in press. Littoral adaptation at the Beagle Channel and surrounding region. In: **Tierra del Fuego: settlement and subsistence on Mankind's southern frontier**, L.A. Borrero and David E. Stuart(eds.). Academic Press.

Orquera, L.A., Sala, A.E., Piana, E.L. & Tapia, A.H. 1978. **"Lancha Packewaia: arqueología de los canales fueguinos".** Huemul (ed.), Buenos Aires:

Orquera, L.A., Piana, E.L. & Tapia, A.H. 1984. Evolución adaptativa humana en la región del Canal Beagle. I-II-III. **Primeras Jornadas de Arqueología de la Patagonia (Trelew).**

Ortiz Troncoso, O.R. 1975. Los yacimientos de Punta Santa Ana y Bahía Buena (Patagonia austral): excavaciones y fechados radiocarbónicos. **Anales del Instituto de la Patagonia**, VI: 93-122, Punta Arenas, (Chile).

Ortiz Troncoso, O.R. 1979. Nuevo fechado radiocarbónico para la Isla Englefield (seno Otway, Patagonia austral). **Relaciones de la Sociedad Argentina de Antropologia, XII** (1978): 243-244.

Ortiz Troncoso, O.R. 1980. Punta Santa Ana y Bahía Buena: deux gisements sur une ancienne ligne de rivage dans le detroit de Magellan. **Journal de la Societe des Americanistes,** LXVI:133-204. Paris.

Osborn, A.J. 1977. Strandloopers, mermaids and other fairy tales: ecological determinants of marine resources utilization - the Peruvian case. In: **For theory building in archaeology,** L.R. Binford (ed.), 157-205. Academic Press.

Panarello, H. 1986. Oxygen-18 temperatures on present and fossil invertebrate shells from Túnel site, Beagle Channel, Argentine. In: **Quaternary of South America and Antarctic Peninsula.** Vol. 5:85-94.

Perlman, S.M. 1980. An optimum diet model, coastal variability and hunter-gatherer behavior. In: **Advances in archaeological method and theory,** M.B. Schiffer (ed.), 3:257-310. Academic Press.

Piana, E.L. 1984. Arrinconamiento o adaptación en Tierra del Fuego. In: **Ensayos de Antropología argentina año** 1984:7-110. Editorial de Belgrano, Buenos Aires.

Piana, E.L. & Orquera, L.A. 1985. Octava campaña arqueológica en Tierra del Fuego: la localidad Shumakush. **VIII° Congreso Nacional de Arqueología Argentina** (Concordia, Argentina).

Polach, H.A. and Golson, J. 1968. Recolección de especímenes para datación radiocarbónica e interpretación de los resultados. **Museo Etnográfico Municipal Dámaso Arce, Monografía 3,** Olavarría (Argentina).

Porter, S.C.; Stuiver, M. & Heusser, C.J. 1984. Holocene sea-level changes along the Strait of Magellan and Beagle Channel (southernmost South America). **Quaternary Research,** 22(1):59-67.

Rabassa, J.; Heusser, C.J. & Stuckenrath, R. 1986. New data on Holocene sea transgression in the Beagle Channel: Tierra de Fuego. In: **International Symposium on sea level changes and Quaternary shorelines,** São Paulo (Brasil). Abstracts.

Rogers, T. 1980. Answering "Maritime-hunter-gatherers: ecology and prehistory". **Current Anthropology** 21(6):727-750.

Sander, F. & Steven, D.M. 1973. Organic productivity of inshore and offshore waters of Barbados: a study of the island mass effect. **Bulletin of Marine Science** 23:771-792.

Saver, C.O. 1962. Seashore: Primitive home of man? **Proceedings of the American Philosophical Society.** 106:41-47.

Saxon, E.C. 1979. Natural Prehistory: the archaeology of Fuego-Patagonia ecology. **Quaternaria** XXI:329-356.

Shepard, F.P.; Curray, J.R.; Newman, W.A.; Bloom, A.L.; Newell, N.D.; Tracey, J.I. & Vee, H.H. 1967. Holocene changes in sea-level: evidence in Micronesia. **Science,** 157:542-544.

Spegazzini, 1982. Costumbres de los habitantes de la Tierra del Fuego. **Anales de la Sociedad Científica Argentina.** XIV: 159-181. Buenos Aires.

Thoms, B.G.; Hails, J.R. and Martin, A.R.N. 1969. Radiocarbon evidence against higher postglacial sea-levels in eastern Australia. **Marine Geology,** 7(2):161-168.

Urien, C.M. 1966. Edad de algunas playas elevadas en la península de Ushuaia y su relación con el ascenso costero postglaciario. **Actas de las Terceras Jornadas Geológicas argentinas** (Comodoro Rivadavia, Chubut), II:35-41.

Weddell, J. 1825. **A voyage towards the South Pole performed in the years 1822-1824 by James Weddell, Master in the Royal Navy.** Loneman, Hurst, Rees, Orne, Brown and Green, London.

Yesner, D.R. 1980. Maritime hunter-gatherers: ecology and prehistory. **Current Anthropology,** 21(6):727-750.

Yesner, D.R. 1983. **Life in the Garden of Eden: constraints of marine diets for human society.** Wenner Gren Foundation for Anthropological Research International Symposium (Cedar Cove, Florida), simposio 94: Food preferences and adversions.

Selected papers of the special session on
the Quaternary of South America,
XIIth INQUA International Congress,
Ottawa, 31 July-9 August 1987

Edited by
JORGE RABASSA
CADIC-CONICET, Ushuaia, Tierra del Fuego, Argentina
KENITIRO SUGUÍO
Instituto de Geociências, Universidade Federal de São Paulo, Brazil

CHALMERS M.CLAPPERTON
Department of Geography, University of Aberdeen, Scotland, UK

10

Maximal extent of late Wisconsin glaciation in the Ecuadorian Andes

ABSTRACT

Sediments deposited during various glacial stages in the Ecuadorian Andes can be broadly differentiated on the basis of altitude, morphology, stratigraphy, degree of weathering and, in places, radiocarbon dating. Massive moraines commonly mark the maximal extent of the last (Wisconsin) glaciation, terminating at a mean altitude of c 3 600 m. Immediately within the end moraines three to four smaller moraines are clustered closely together, indicating subsequent stadial or readvance positions of the glaciers. Distinctive moraine systems at altitudes of c 3 990 - 4 100 m and c 4 200 - 4 400 m date from the Late-glacial (c 10 000 - 12 000 BP) and Neoglacial (c 5 000 BP - late 19th. C) periods respectively.

Six radiocarbon dates from organic sediments interbedded with tills indicate that glaciers in Ecuador expanded after c 33 000 BP and that the organic sediments had been accumulating since before c 43 000 BP. Three dates between 33 000 BP and 38 000 BP from the uppermost organic layers suggest that Ecuadorian glaciers may have stood at their maximal late Wisconsin limits well before the northern ice sheets reached theirs. The glaciers shrank from these limits during the last global maximal glaciation (at c 18 000 BP) because of increased aridity, but fluctuations occurred before they finally receded sometime after c 15 000 BP.

RESUMEN

Los sedimentos depositados durante varios estadíos glaciales

en los Andes Ecuatorianos pueden ser diferenciados en general
sobre la base de su altitud, morfología, grado de meteoriza-
ción y, en algunos lugares, dataciones radiocarbónicas. Morenas masivas marcan por lo común la extensión máxima de la
última glaciación (Wisconsin), culminando a una elevación
promedio de ca. 3600 m. Inmediatamente dentro de las morenas
terminales, tres o cuatro morenas más pequeñas están agrupadas
muy cerca unas de otras, indicando posiciones de reavance o
estadiales de estos glaciales. Sistemas morénicos distintivos
a alturas de ca. 3990 - 4100 m y ca. 4200 - 4400 m datan del
Glacial tardío (ca. 10000 - 12000 a AP) y el Neoglacial
(ca 5000 AP a fines del siglo XIX), respectivamente.

Seis dataciones radiocarbónicas de sedimentos orgánicos
interestratificados con tills indican que los glaciares en
Ecuador se expandieron luego de ca. 33000 AP y que los sedimentos orgánicos habían sido acumulados desde antes de
ca. 43000 AP. Tres fechadas entre 33000 AP y 38000 AP,
provenientes de los niveles orgánicos superiores, sugieren
que los glaciares ecuatorianos podrían haber alcanzado sus
límites máximos para el Wisconsin tardío mucho antes que las
calotas continentales del Hemisferio Norte llegaran a los
suyos propios. Los glaciares retrocedieron desde estos límites
durante la última glaciación global máxima (hacia aproximadamente 18000 AP) a causa del incremento de la aridez, pero
algunas fluctuaciones tuvieron lugar antes que ellos final-
mente se retiraran en algún momento luego de ca. 15000 AP.

INTRODUCTION

A sequence of late Quaternary glaciation in the Ecuadorian
Andes has recently proposed following a reinterpretation of
earlier work by Walter Sauer (Sauer 1965, Clapperton and Vera
1986). Four main stages are now identified (Clapperton
1986). The oldest is represented by deeply weathered and
highly oxidised till and reaches its lowest altitudinal
limits of c 2750 m in the eastern and southern cordilleras;
the degree of weathering and absence of clear morainic
landforms suggest that this deposit is older than the last
glaciation. The second oldest is normally marked by very large
lateral-terminal moraines extending to their lowermost limits
of c 3000-3600 m on the eastern slopes of the cordilleras.
This stage is frequently characterised by the presence of

three-four smaller moraines closely spaced together a short
distance inside the end moraine limits; it has been assumed
that these deposits formed during the last glaciation. The
two younger moraine stages, typically terminating at c 4000
and 4250 m, are reasonably well dated to the Late-glacial and
Neoglacial periods with ages 12,000-10,000 BP and since
c 5000 BP respectively (Clapperton and McEwan 1985, Clapperton
1986).

The maximal extent of glaciers during the last glaciation
is normally assumed to have occurred in most parts of the
world at c 18,000 BP (Denton and Hughes, 1981). However,
Porter (1981) has shown that western outlets from Andean
icefields formerly covering the Chilean Lake District
(lat 40 -42°S) reached their greatest extent between 29,600 BP
and 21,100 BP, and Van der Hammen et al. (1981) considered
that during the later part of the last glaciation glaciers
in the Sierra Nevada del Cocuy, Colombia, were at their outermost limits between c 28,000 EP and 25,000 BP. This paper
presents results from the radiocarbon dating of organic
sediments interbedded with glacial deposits in Ecuador in an
attempt to establish a more precise chronology for the last
(Wisconsin) glaciation in the tropical Andes. The three sites
investigated are in the Carihuairazo massif in the western
cordillera, the Laguna Pisayambo area of the eastern cordillera and Quebrada Tomebamba in the southern cordillera (Figure
1).

THE CARIHUAIRAZO MASSIF

Nevado Carihuairazo (5,020 m) is a deeply dissected extinct
volcano which presently supports nine small glaciers despite
the relatively low altitude of the mountain crests (Figure 2).
Sections exposed by road cuts, streams and glacial activity
show that the volcano is composed predominantly of andesitic
lavas interbedded in places with minor amounts of tephras and
other volcanic deposits. Hall (1977) noted the presence of a
remnant caldera c 2 km in diameter open towards the north and
north-east. Debris avalanche deposits more than 15 m thick,
together with pyroclastic flow units, lahar deposits and
tephras cropping out on the lower north-eastern flanks of the
volcano appear to represent late paroxysmal eruptions which
may have formed the caldera-like feature. Aspects of the

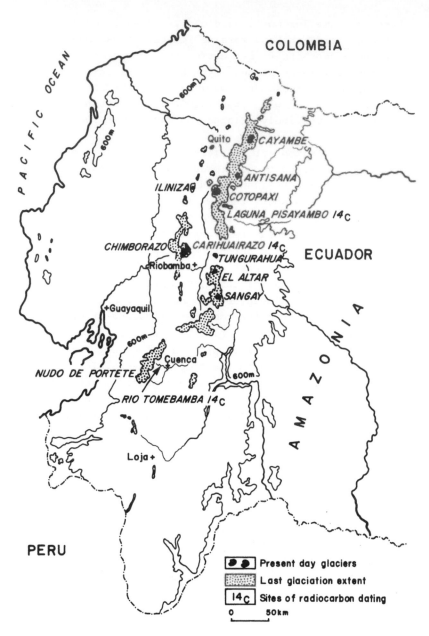

Figure 1. Extent of glaciers in the Ecuadorian Andes during the maximal stage of the last glaciation; radiocarbon dated sites are indicated.

glacial sequence for the combined Chimborazo-Càrihuairazo
massif have been discussed by Clapperton and McEwan (1985)
and Clapperton (1986), and the former extent of ice fields
on Carihuairazo during the Necglacial, Late-glacial and
Full glacial periods is shown on Figure 2. This paper is
concerned only with the glacial-volcanic stratigraphy exposed
on the north and north-east flanks of the mountain and on the
radiocarbon dating of interbedded organic sediments.

Figure 3 is a composite diagram illustrating a generalised
interpretation of the probable stratigraphy underlying north-
east slopes of the mountain below an altitude of 4000 m. The
base of the sequence is composed of weathered volcanic
sediments, commonly inclined at angles different from those
of the present slopes. These are overlain by a deeply
weathered diamict at least 4 m thick. This sediment is unsorted
and multilithic, contains clasts of mixed sizes up to 2 m
diameter, many of which are faceted and held in a silty-clay
matrix. These characteristics suggest that the diamict is till
rather than a debris avalanche deposit. The overlying unit
consists of 5-8 m of tephras varying in calibre from fine
sand to 2 cm pumice clasts. In places the beds are gently
inclined in harmony with existing topography, but at one site
they have been deformed by compressive stress, possibly
imparted by an overriding mass of glacier ice; this
interpretation is supported by the presence of glacial deposits
(unit 4) overlying the tephras at some sites. The three lower
units (1-3, Figure 3) typically crop out at the ends of
interfluves between major valleys, from which they are
generally absent; this implies that they may pre-date the
development of the main valleys. Unit 4 consists of till, the
upper few metres of which have been oxidised; clasts retaining
striations are nevertheless abundant. An attempt was made
to obtain a statistical measure of the thickness of weathering
rinds on stones c 5-15 cm in size but proved unsuccessful
because of the large variety of lithologies present. Typically,
coarse-grained clasts are weathered throughout whereas fine-
grained lithologies have weathering rinds 4-6 mm in thickness.
In places, unit 4 till forms the core and outer slopes of
large moraine ridges located within the major valleys, and
was clearly deposited after the valleys had been cut. Only
one or two thin tephra layers appear to be present on top of
this morainic deposit. Unit 5 consists of interbedded silts,
sands, gravels and peat beds (Figure 4); the maximum observed

169

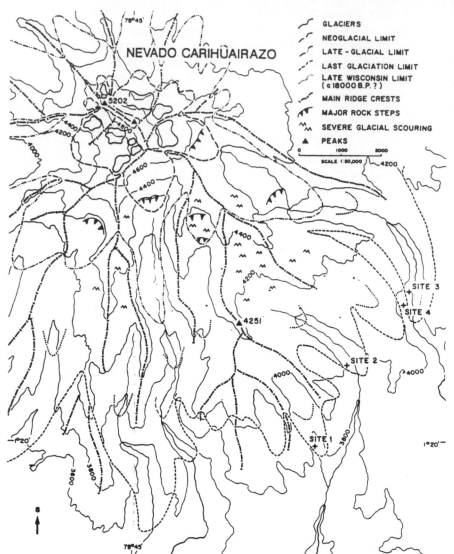

Figure 2. Limits of the Neoglacial, Late-glacial and Wisconsin glacial stages on the northern slopes of Nevado Carihuairazo; dated sites are also shown.

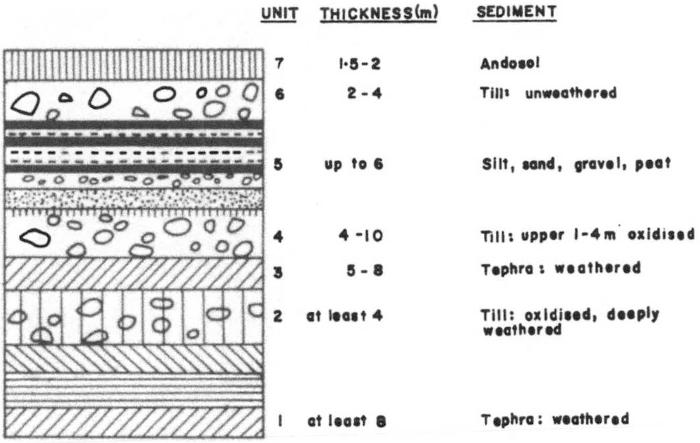

UNIT	THICKNESS(m)	SEDIMENT
7	1·5 - 2	Andosol
6	2 - 4	Till: unweathered
5	up to 6	Silt, sand, gravel, peat
4	4 - 10	Till: upper 1-4 m oxidised
3	5 - 8	Tephra: weathered
2	at least 4	Till: oxidised, deeply weathered
1	at least 8	Tephra: weathered

Figure 3. Composite diagram illustrating stratigraphical relationships interpreted from several sections on the north and north-east slopes of Carihuairazo volcano, Ecuador.

thickness is 6 m. The unit is best preserved on a broad interfluve situated between two major valleys and which appears to have been overridden by relatively thin local glaciers (Figure 2). Sands and gravels interbedded with peat beds suggest the former occurrence of periods of meltwater discharge, most probably associated with enlarged local glaciers. Pockets of disturbed and incorporated peat occur also within some of the large moraines in the major valleys, suggesting that a peat-covered landscape was overridden by an expansion of glaciers down these valleys. Unit 6 comprises relatively fresh, unweathered till that forms a 2-4 m cap overlying the organic and fluvial/glacio-fluvial sediments (Figure 4). It makes up the sharp crests and inner slopes of the large, mcrphologically fresh lateral moraines in the major valleys. This unit is consistently capped by a 1-2 m layer of black andosol, a Holocene aeolian-volcanic soil which is probably still forming in the cool, humid climate of the **paramo** grassland.

Figure 5 shows the stratigraphical details of sites where radiocarbon dates have been obtained from organic sediments comprising unit 5. Implications of the stratigraphy include the following. Since the age span of the organic and fluvial/ glacio-fluvial sediments composing unit 5 is from c 33,000 BP

Figure 4. Details of site 3a (Figure 5) on Carihuairazo, showing fresh morainic till of the last glacial maximum overlying interbedded silts, gravels and peats.

to before 43,440 BP, they most probably accumulated during
a mid-Wisconsin interstadial period. This implies that the
underlying oxidised till of unit 4 may be from an earlier part
of the last glaciation and that the glaciers at that time were
as or more extensive than during the later part of the last
glaciation. The tephras of unit 3 may represent eruptions of
Carihuairazo during the last interglaciation and the deeply
weathered till of unit 2 is probably from the penultimate or
an earlier glaciation. Figure 5D indicates not only that
glaciers had receded from the site by c 15,000 BP but also
that the last major eruption in the area, probably of
Chimborazo volcano, occurred shortly after c 11,400 BP (cf.
Clapperton and Smyth, 1987).

LAGUNA PISAYAMBO

Laguna Pisayambo occupies a shallow bedrock basin enclosed by
massive lateral moraines on the north-western fringe of the
Jaramillo-Archiriqui massif, part of the Llanganate mountains
in the eastern cordillera (Figure 1). The valley in which
the lake is situated is mainly formed by a composite lateral
moraine 150-200 m high and heads some 7-8 km to the south-
east in rugged uplands rising just over 4,200 m in altitude
(Figure 6). There are no glaciers or permanent snowfields in
the area but perennial snowbeds may have existed on occasions
during the last few centuries (Hastenrath, 1981, p.38-39).
The presence of cirques clearly scoured by an all-enveloping
ice cap indicates that the area has undergone local and ice
cap phases of glaciation. The large moraine system surrounding
Laguna Pisayambo marks limits attained by one of the larger
outlet glaciers from a former ice cap and shows that it
terminated at an altitude of 3500 m. Within the last 3 km of
the moraine ridges only a few metres high curve across the
valley floor and can be traced into narrow lateral ridges
on the upper slopes of the massive composite moraine (Figure
6). The level of Laguna Pisayambo has recently been raised
artificially with an earth dam constructed with material
excavated from the lateral moraine bounding the lower part
of the eastern side of the valley. The fresh section (1985)
exposes c 160 m of morainic sediment and the stratigraphy
shown in Figure 7. The internal structure suggests that an
original moraine composing approximately two thirds of the

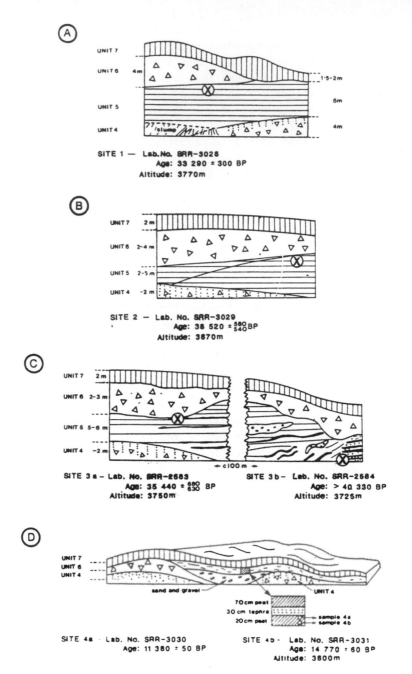

Figure 5. Details of sections exposed on the north and northeast slopes of Carihuairazo volcano, indicating radiocarbon ages of organic sediments. Sampled points are also indicated.

Figure 6. Vertical air photograph of the Laguna Pisayambo
moraine system, Llanganate mountains of the eastern
cordillera. The location of Figure 7 is indicated.

entire deposit was subsequently buried by a younger readvance
till. The upper 30 cm of the lower deposit is slightly oxidised
and has a shallow lens of organic silts preserved in hollows
on the former land surface. The bulk of the large lateral
moraine complex therefore appears to have been formed by two
glacier advances reaching roughly similar limits. Radiocarbon
dating of the buried organic silts has given an age of
34,200+1080 (SRR-3034), indicating an interstadial origin for
950

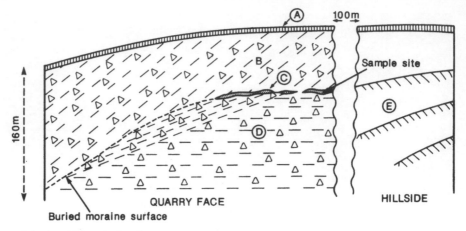

Figure 7. Stratigraphy exposed in a quarry cut into the eastern lateral moraine at Laguna Pisayambo.

Unit A = 1.75m Andosol
* B = 50-160m Moraine:faint,but distinct layering evident
* C = 0-95m Silt and peat
* D = 10-100+m Moraine:v. slight weathering of upper 30cm
* E = Inner slope of lateral deposit with 3 minor lateral ridge crests

Lab. No. SRR-3078

Age. 34200 $^{+1080}_{-950}$ B.P.

Altitude. 3550m

the material. The lower moraine deposit may, therefore, have formed during an earlier part of the last (Wisconsin) glaciation whereas the overlying moraine dates from a later stage.

QUEBRADA TOMEBAMBA

Quebrada Tomebamba drains south-eastwards from the rugged plateau composing the Nudo de Portete massif in the southern cordillera of Ecuador (Figure 1). As in the Llanganate mountains, glaciers are presently absent because the highest ground rises only to 4300 m whereas the regional snowline is probably about 4800 m. However, the widespread presence of ice-scoured surfaces on the plateau testifies to the former existence of a large ice cap during the last glaciation (Clapperton 1983, Plate 2(a)). Outlet glaciers from this

176

catchment area spread down major valleys such as that of the
Río Tomebamba (Figure 1) and deposited conspicuous moraines
at c 3000 m altitude. A road cut approximately 2 km down-
valley from these limits exposes c 25 m of deeply weathered
till overlain by an organic-rich palaeosol 20-30 cm thick
which, in turn, is buried by 1.5-2 m of soliflucted debris.
The palaeosol gave an indeterminate radiocarbon age in excess
of 43,110 BP (SRR-2588). The significance of this site is
difficult to assess, but in view of the limited depth of soil
development and absence of any tephras, the soil may have
formed during an interstadial period roughly mid-way through
the last (Wisconsin) glaciation. The overlying solifluction
material in this case would correspond to the cold later part
of the last glaciation.

DISCUSSION

Radiocarbon dates from the uppermost layers of organic
sediments resting on oxidised till and buried by fresh
morainic deposits at three sites in the Ecuadorian Andes lie
within the range c 33,000-38,000 BP. Three main implications
arise from this evidence. One is that glaciers readvanced
after a distinct interstadial period that began sometime
before c 43,000 BP. Although the length of this period has not
yet been closely bracketed, it may well correspond to the
mid-Wisconsin/Weichselian interstadial, identified from
biostratigraphical and oxygen isotope studies, with an age
range of c 30,000-50,000 BP (e.g. Lowe and Walker, 1984
p.308-310). A second is that glaciers advanced slightly
farther down-valley before the interstadial than they did
afterwards; the relatively restricted amount of oxidation and
weathering of the pre-interstadial till (unit 4) supports
the interpretation that the till may date from an early part
of the last glaciation rather than from the penultimate
glaciation. A third is that during the later part of the last
glaciation, Ecuadorian glaciers probably reached their
greatest extent well before the recognized global maximum of
glacial expansion at c 18,000 BP. Although some organic
material may have been removed by overriding ice, the
youngest date of c 33,000 BP suggests that the glaciers may
have culminated their advance no more than a few thousand
years later. It is interesting to note that glaciers in the
adjacent tropical Andes of Colombia and in the temperate

Andes of Southern Chile appear to have achieved their maximal late Wisconsin extent between c 25,000 and 29,000 BP (Van der Hammen **et al.** 1981, Porter 1981). If there has been a measure of synchrony in major periods of glacier expansion throughout the Andes, then it seems likely that Ecuadorian glaciers would have culminated their late Wisconsin/Weichselian advance also during this period of time. Porter's work in the Llanquihue district of Chile has shown that the large outlet glaciers fluctuated close to their maximal position until about 13,000 BP, although they became slightly reduced in extent after c 19,000 BP. Similarly, in Colombia, interpretation of palynological and glacial geological data led Van der Hammen **et al.** (1981) to conclude that glaciers receded within their earlier end moraine limits as the climate became colder but much drier. Increased aridity in tropical South America during the last glacial maximum has long been suspected from evidence such as relict sand dune fields in the Orinoco basin (Khobsi, 1981) and biological refugia in Amazonia (Colinvaux, 1979).

This paper concludes that glaciers in the Ecuadorian Andes may have culminated their late Wisconsin/Weichselian advance before c 25,000 BP and remained close to, but within, their maximal limits until at least 15,000 BP. Fluctuations of the glaciers during this time produced the morainic landforms and sediments located within the end moraine limits.

ACKNOWLEDGEMENTS

Thanks are due to the Royal Society, The Carnegie Trust for the Universities of Scotland, and Aberdeen University for supporting fieldwork in Ecuador in 1983 and 1985. I am also most grateful to the Natural Environmental Research Council and, in particular, Dr D.D. Harkness for the radiocarbon dating assays.

REFERENCES

Clapperton, C.M. 1983. The glaciation of the Andes. **Quaternary Science Reviews,** 2:81-155.
Clapperton, C.M. 1986. Glacial geomorphology, Quaternary glacial sequence and palaeoclimatic inferences in the

Ecuadorian Andes. Proceedings of First International
Conference on Geomorphology, Manchester 1985, 2:770-843.

Clapperton, C.M. & McEwan, C. 1985. Late Quaternary moraines
in the Chimborazo area, Ecuador. **Arctic and Alpine Research,**
17:135-142.

Clapperton, C.M. & Smyth, M.A. 1987. Late Quaternary debris
avalanche from Chimborazo volcano, Ecuador. **Memoria del
Simposio Sobre Neotectónica y Riesgos volcánicos,** Bogotá,
1986. (in press).

Clapperton, C.M. & Vera, R. 1986. The Quaternary glacial
sequence in Ecuador. A reinterpretation of the work of
W. Sauer. **Journal of Quaternary Science,** 1:45-56.

Colinvaux, P. 1979. The Ice-age Amazon. **Nature** 278:399-400.

Denton, G.H. & Hughes, T. 1981. The Last Great Ice Sheets.
Wiley-Interscience, 484pp. New York.

Hall, M. 1977. El volcanismo en el Ecuador. Instituto Pana-
mericano de Geografía e Historia, 120pp. Quito.

Hastenrath, S. 1981. The Glaciation of the Ecuadorian Andes.
159pp, Rotterdam: Balkema.

Khobsi, J. 1981. Los campos de dunas del norte de Colombia y
los Llanos del Orinoco (Colombia y Venezuela) In, Memoria
del primer Seminario sobre el Cuaternario de Colombia,
Bogotá. **Revista CIAF,** 6:257-292.

Lowe, J.J. & Walker, J.C. 1984. Reconstructing Quaternary
Environments. 389pp. London: Longman.

Porter, S.C. 1981. Pleistocene glaciation in the southern Lake
District of Chile. **Quaternary Research,** 16:263-292.

Sauer, W. 1965. Geología del Ecuador. 383pp. Quito: Editorial
del Ministerio de Educación.

Van der Hammen, T., Barelds, J., De Jong, H. & De Veer, A.A.
1981. Glacial sequence and environmental history in the
Sierra Nevada del Cocuy. (Colombia). **Palaeogeography,
Palaeoclimatology, Palaeoecology.** 32:247-340.

CLAUDIO A.SYLWAN

Geological Institute, University of Stockholm, Sweden

11

Annual paleomagnetic record from late glacial varves in Lago Buenos Aires Valley, Patagonia, Argentina

ABSTRACT

A sequence of 887 varves has been measured. A continuous process of ice retreat is inferred through the gradual decreasing of the thickness of the varves. Paleomagnetic studies show very strong intensities and stable polarity within single varves as well as from varve to varve. The general trend of declination is a smooth shift from 40-50°W to 0° at the top. The inclination shifts from -70°-85° to -35° at the top. The VGP is localized in the Arctic-Siberian region and swings later to western Greenland-northeastern Canada. The age of the varves is on the range between 12,500-13,500 BP.

RESUMEN

Una secuencia de 887 varves ha sido medida. Un proceso continuo de recesión glacial puede inferirse a través de la disminución gradual del espesor de los varves. Los estudios paleomagnéticos muestran intensidades muy fuertes y polaridad estable en varves individuales, así como de varve a varve. La tendencia general de la declinación es una variación suave de 40°-50°W, a 0° en el techo de la secuencia. La inclinación varía de -70°-85° a -35° en el techo. El PGV está localizado en la región ártico-siberiana y se desplaza luego hacia Groenlandia occidental o Canadá nororiental. La edad de lcs varves parece encontrarse en el rango entre 12.500-13.500 años AP.

INTRODUCTION

In 1932, Caldenius published an extensive synthesis of the
Quaternary glaciations in Patagonia. He mapped the succession
of moraine ridges all along the eastern side of the major ice
cap in southern South America. He also measured the numbers
of and variations in thickness of late glacial varves in a
number of profiles. Because of the presumed correlation with
the Fennoscandian varve record and moraine succession, he
assigned an age of the varves of about 10,000 BP.

In the Lago Buenos Aires region (46° 35' S, 71°W), Caldenius
(1932) reported the finding of 784 annual varves. We chose
this locality for detailed paleomagnetic analyses because
the number of varves was so large that it seemed possible
to obtain statistically significant correlations with the
paleomagnetic changes in varve records (Mörner, 1982) and/or
other high-sedimentation-rate sequences (e.g. Easterbrook
et al., 1976; Creer, 1977; Thouveny **et al.**, 1985) of similar
age from the northern hemisphere. This was done in order to
see if it was possible to descipher any paleo-dipole movements
or just paleo-non-dipole changes. At the same time, there
are only a few works on Quaternary paleomagnetism in Patagonia
on postglacial lake sediments (e.g. Valencio **et al.**, 1982;
Sinito **et al.**, 1985) or glacial sediments (e.g. Kodama **et al.**,
1985). Therefore, our investigations in the Lago Buenos Aires
area have also a more general paleomagnetic and geologic
interest.

This is the first integrated analysis of the varve
chronology and paleomagnetism in Quaternary varved sedimentary
sequences on the southern hemisphere. A detailed report on
the glaciological and sedimentological aspects will be
presented elsewhere, a preliminar report on the long termed
glacial chronology is given by Mörner and Sylwan (1987).

SAMPLING PROCEDURES

Outcroping, horizontally bedded, late glacial varves in the
Río Fenix Chico (Figure 1) were measured and sampled in two
vertical profiles (labelled "A" and "B"). Oriented samples
were collected in the field in cubic plastic boxes (side:
2.1 cm), either by gently driving the boxes into the wet,
silty and clayey sediments or by carving the samples to fit
the boxes when the field moisture conditions were not

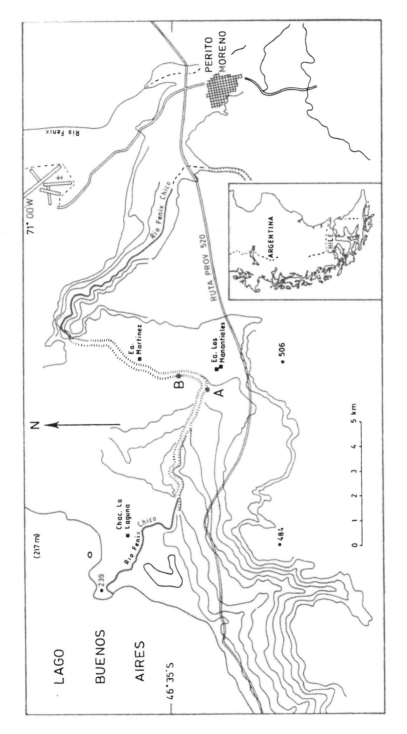

Figure 1. Río Fenix Chico, sites of the paleomagnetic samplings (profiles "A" and "B").

suitable for pushing the boxes into the sediments. All these
procedures were done without affecting the sediment structures.
During sampling, the orientation of each container was
maintained through careful alignment to the magnetic north and
to the horizontal. Once the sample was taken, the orientation
was marked directly on the top face of the cube, which later
was sealed with non-magnetic glue to keep the sample intact
and to avoid the shrinkage of the sediment through evaporation.
Due to the high time resolution, only one sample was - as a
rule - taken per varve. In some varves, however, several
samples were taken. In a few cases, samples were taken by
different techniques (plastic boxes and small piston cores)
and from perpendicular sides in order to have a check on
possible mechanical effects. Profile "A" is 7.20 m high and
contains 60 varves, they represent the base of the whole
sequence, from them 59 varves were taken for paleomagnetic
studies. The main profile "B", is 35 m high and contains
827 varves. The paleomagnetic sampling extends from varve 1
to varve 600, the subsequent (227) varves, being deformed
and unsuitable for paleomagnetic analyses.

STRATIGRAPHY

In combination, the two profiles from Río Fenix Chico form
a 42 m sedimentary sequence covering 887 varves. The thickness
of the varves were carefully measured as well as the thickness
of five interbedded till layers in the lower part of the
sequence. The thickness of the varves diminishes upwards in
a way which clearly speaks for a successive (unbroken) ice-
retreat. Figure 2 shows the mean values of the size of the
varves versus time. A curve (plotting varve thickness against
time), of very similar slope was established by Ignatius
(1958) for the Baltic region and used as evidence of a single
and continuous process of ice-melting.

  Sedimentologically, the clastic composition of the varves
varies through the sequence. The varves of Río Fenix Chico
have a very marked diatactic character consisting of a couplet
of contrasting laminae which represent the seasonal
sedimentation within a year; coarser and lighter summer units
and finer and darker winter units. In general, a relationship
between the thickness of the varves and the clastic size of
their material is distinct; both thickness of varves and
grain size diminish upwards. This fact is also verified in

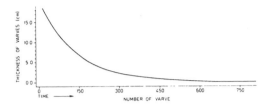

Figure 2. Mean thickness of the varves from Río Fenix Chico.

the irregularly large varves that lie scatteredly along the
succession. Bottom varves, the mean thickness of which is
about 10-15 cm, are composed of silty sand and sandy silt in
the summer part, while the winter part is composed of clayey
silt. This composition varies gradually to reach the top of
the series with silt and clay in the summer part and with
clay in the winter one. In some cases, even sand and coarse
silt conform the different parts. The proportion between the
summer and the winter parts within a varve also varies through
the succession. At the bottom, winter parts attain a mean
percentage of 15% of the total varve thickness, reaching
- in extreme cases - from 10 to 30%. Upwards, this proportion
changes gradually to 50% at the top.

Varve 343 is very interesting because it contains a 5 cm
layer of green volcanic ash which overlies a horizon with
very abundant carbonatic concretions. Other ash layers have
been found in varves 493 (0.3 cm of yellow ash), 563 (1.2 cm
white silty ash), 586 (0.3 cm yellow ash), 640 (0.2 cm yellow
ash) and 708 (very thin layer of yellow ash). Varve 219 is
21.4 cm thick and contains large amounts of gipsy crystals
mixed with brown sand.

Five till beds are interbedded in the sequence at the place
where the profiles were carried out (Figure 7). These lens-
shaped deposits of till seem only to be of local importance,
trending to disappear laterally. Tills are mainly composed
of poorly sorted material, e.g. coarse silt, fine, middle
and coarse sand and pebbles of about 1-2 cm which appear
scatteredly. The till beds or flow tills are interpreted not
representing glacial readvances but only material slided
off/melted out from the ice cap or, rather, from drifting
ice bergs. This is supported by the character of the varves
just below and just above the till beds.

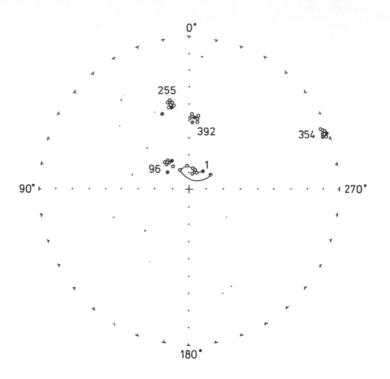

Figure 3a. AF demagnetization paths for five specimens on stereonet. Steps: NRM - 100 - 200 - 300 - 450 - 600 - 800 and 1000 Oe.

PALEOMAGNETIC MEASUREMENTS

The samples were all measured at the Stockholm Paleomagnetic laboratory on a Digico magnetometer system and cleaned on an alternating field demagnetizer.

Susceptibility measurements were carried out on a bulk susceptibility unit. Except for a few samples the natural remanent magnetization (NRM) and the remanent magnetization remaining after 200 Oersted alternating field demagnetization, were measured. 16 samples were demagnetized at 50, 100, 200, 300, 450, 600, 800 and 1,000 Oersted intervals. Figures 3a

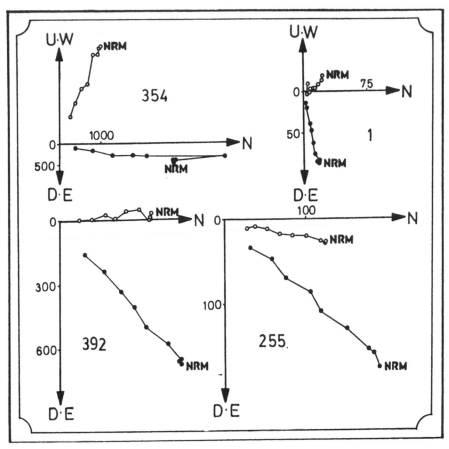

Figure 3b. Demagnetization diagram according to Zijderveld (1967) for four specimens at NRM - 50 - 100 - 200 - 300 - 450 - 600 - 800 and 1000 Oe.

and 3b show that demagnetization has generally a very small effect on the NRM directional values. Thus, the 200 Oersted demagnetization was chosen as the best standard peak field and applied to all the samples.

SUSCEPTIBILITY AND INTENSITY

Diagrams for susceptibility and intensity are shown in Figures 4 and 5. As intensity of remanence is dependent on the proportion of the magnetic minerals, this abundance was

measured in terms of magnetic susceptibility. Susceptibility
shows stable values throughout the entire sequence. Lowest
value (around 20   G/Oe) are encountered scattered through
the series and mostly belong to varves which are much wider
than the mean thickness, although not everyone that is wider
than the mean value shows low intensity. Susceptibilities lie
mainly in the range of 80-200   G/Oe, reaching 265   G/Oe in
the lower part of the sequence (bottom of profile "A"). The
shapes of intensity and susceptibility records are quite
similar and there is a good agreement between the peaks of
low values in both curves. The Q factor (intensity/
susceptibility) swings smoothly to higher values upwards in
the sequence, where the thickness of the varves has its
minimum. This shift is from about 1-2 to about 6-7 and should,
in principle, be related to the paleofield intensity.

DIRECTIONAL PALEOMAGNETIC VARIATIONS

The declinations and inclinations from Río Fenix Chico are
plotted in Figures 6 and 7. The declination record shows a
general trend which smoothly shifts from 40°-50° at the base
to around 0° at the top of the sequence. Scatter of about 60°
amplitude are observed at the profile "A" and in the lower
part of "B". A marked oscillation which reaches a value of
100° to the east is observed between varve 70 and 90. Above
this oscillation the values become more stable and, except
for a few scattered values of westerly declination, the record
lies around 0° up to the top.

The inclination record is characterized by a general
shallowing from values of about -70° and 85° at the base to
values of about -35° at the top. All the inclination values
measured were negative. In particular, it is noticeable that
values from varves below and above till beds present acceptable
agreements. There is no reason to discuss the minor shorter
scattered of the individual samples, the significant message
is the long term trend.

Virtual Geomagnetic Pole positions (VGP) were calculated for
all samples (Figure 8). Three main positions can be observed
(Sylwan, 1985). The lower part of the series (profile "A")
gives a position in eastern Siberia. The VGP of the top of "A"
and the bottom of "B" is fairly scattered with a more stable
position (at the top) in around northwestern Siberia. From the

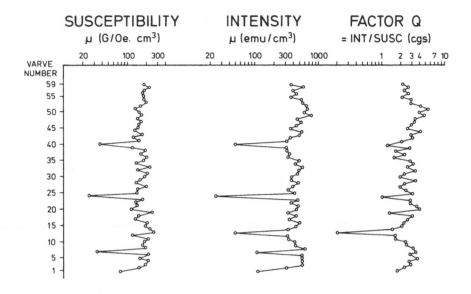

SUSCEPTIBILITY	INTENSITY	FACTOR Q
μ (G/Oe. cm³)	μ (emu/cm³)	= INT/SUSC (cgs)

Figure 4. Susceptibility, intensity and factor Q from profile "A".

middle of profile "B", a gradual shift to the west is recorded, attaining a final position in northeastern Canada-western Greenland.

CHRONOLOGICAL POSITION OF THE VARVE SEQUENCE

Caldenius (1932) assigned a very close age of 10,200-9,500 BP to the varve sequence in question. This was only based on the so-called teleconnections which are in no way tenable.
To obtain a proper direct or indirect dating, we have to turn to radiocarbon dating from the area itself or to the timing of the general climatic changes in the region.
  Mercer (1970, 1972, 1976) showed that there was a major glacial maximum in southern South America at around 19,500 BP with a secondary small readvance at around 13,300 BP, and that "deglaciation..... was very extensive by 12,000 BP" (because the gyttja production has started at that time in the Lago Ranco area). The retreat from the maximum position was 100% already by 11,000 BP according to Mercer (1972, Figure 13).
  The exponential decay of mean varve thickness (Figure 2)

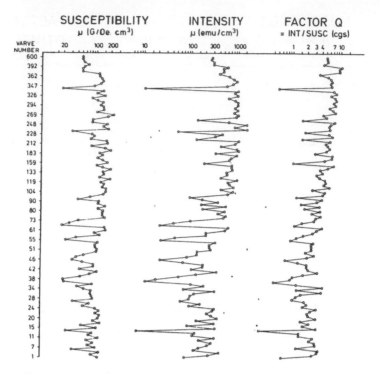

Figure 5. Susceptibility, intensity and factor Q from profile "B".

and the general character of the whole varve sequence as well
as of the individual varves or varve segment seem to indicate
that the ice withdrew without any significant oscillations.
The Lago Buenos Aires ice lake, in which the varves were laid
down, drained when the outlet in the west was free-melted.
The distance from the major moraine just east of the varve
locality to the outlet in the west is about 145 km. If it
took about 890 varves for the ice to recede this distance,
the mean annual rate of recession is about 160-165 m/year.
This fits well with the fact that the scattering of the
individual varves around the mean curve exhibit a significant
change at around varve 375 from a large scattering to little
or no scattering. An ice recession of 165 m/year for 375 years
would bring the ice margin from an open lake position to a
position in the narrow extension on the Chilean side.

Radiocarbon analyses have been done from four varves.
Carbonatic concretions appear sporadically in the sequence.

190

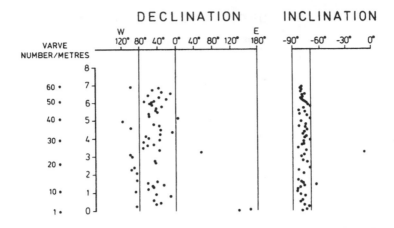

Figure 6. Declination and inclination from profile "A".

In varve 343 there is a high concentration of them, however.
Variation in age of the different fractions of the dated
concretions in varve 343 reveal that they are younger (post-
depositional) diagenetically grown concretions. The
concretion in varve 62 has a C13 value which, in comparison
with the external fraction of the sample from varve 343,
suggests that it has a diagenetic origin, too. There are two
dates left which seem to represent the true age of the varves.
The overlapping time of the two dates is 13,100-13,375 BP
with the corresponding varve difference being 286 years
(Table 1). This suggests that our varve sequence (base of
profile "A") started to accumulate at around 13,450 BP and
ceased at about 12,550 BP (top of profile "B").

DISCUSSION

The results from the paleomagnetic measurements and the VGP
positions obtained from the varves of Río Fenix Chico are
interesting in relation to the age of the ice retreat in the
Lago Buenos Aires Valley.
   As similar records in age and in rate of resolution are
lacking for the Patagonian region, it is not possible, at the
moment, to correlate this records with others. On the other
hand, some similarities with the VGP positions in records from

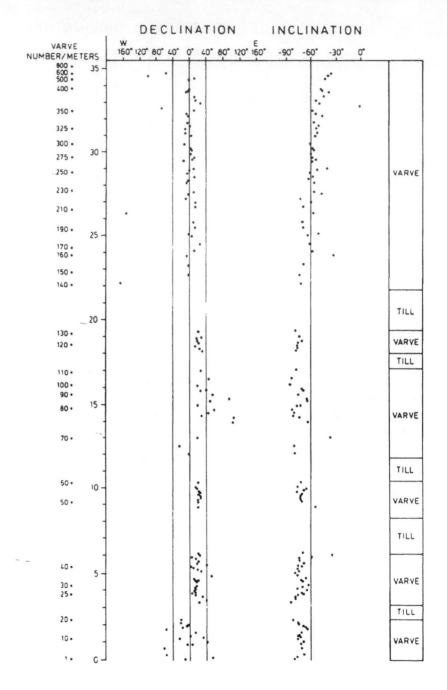

Figure 7. Declination, inclination and stratigraphy from profile "B".

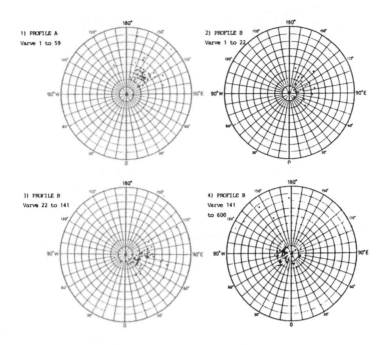

Figure 8. The stereonets show the different VGP positions along the sequence.

Table 1. Comparison of varves and radiocarbon dates from varves 20 and 306 in profile "B".

Varve	Varve Difference	C14 Date	Range with 2	Overlapping Age	Difference in years
306		12,840±130	12,580-13,100	ca. 13,100	
	286				275
20		14,065±345	13,375-14,755	ca. 13,375	

the northern hemisphere have been found. It is worthy to point out the VGP shift from western Siberia to western Greenland-northeastern Canada, which has also been found by Mörner (1982) from varved sediments in Ontario, Canada. The age of this shift is assumed to be 13,200 years BP. A very similar data

193

is described by Easterbrook and Othberg (1976) for a record obtained in Puget Lowland, State of Washington, where the VGP of Vashion sediments (age of the top of the formation: 13,000 BP) shifts from northwestern Siberia to west of Greenland for Kulsham glaciomarine drift (age of the top of the sequence: 11,640 BP). The record obtained by Creer **et al.** (1986) from Lac du Bouchet (Haute Loire, France) also shows a westerly shift of the VGP around the period here considered.

If this is true, it would suggest that we are dealing with short-term changes of the dipole. This is interesting as the Late Quaternary secular paleomagnetic changes are almost entirely interpreted in terms of non-dipole changes (e.g. Creer, 1981).

CONCLUSIONS

We have applied paleomagnetic analyses to 659 varves from a sequence of totally 867 varves in Patagonia, southern South America. The paleomagnetic analyses gave unusually strong and stable values. Demagnetization has little or no effect on the directional components. The general trend is a shift in declination from about 40°-50°W to 360°, and a shift in inclination from about -70°-85° to -35°. The Q-factor exhibits a general shift from about 1-2 to 6-7. The corresponding VGP shift from Siberia to northeastern Canada-western Greenland agrees with other detailed records from the northern hemisphere. Through C14 analyses we have determined the age of the varved sequence to between 13,450 and 12,550 BP.

ACKNOWLEDGEMENTS

The main content of this paper was presented at the IAGA 1985 Conference in Prague (session 01-16).

The field work was supported by the Department of Geology, University of Stockholm, and by separate funds issued by the University of Stockholm. The C14 analyses were made at the Laboratory of Isotopic Geology, National Museum of Natural History, Stockholm.

The author wants to thank Dr. Nils-Axel Mörner for technical assistance and for many helpful comments and suggestions. Thanks are also due to the authorities of the Municipalidad of Perito Moreno for the facilities given during the field work.

REFERENCES

Caldenius, C.C. 1932. Las glaciaciones cuaternarias en la
Patagonia y Tierra del Fuego. **Geogr. Annal.** H. 1 a. 2:1-164.

Creer, K.M. 1977. Geomagnetic secular variations during the
last 25,000 years: an interpretation of data obtained from
rapidly deposited sediments. **Geophys. Jour. Royal Astronom.
Soc. London** 48 (1977):91-109.

Creer, K.M. 1981. Long-period geomagnetic secular variations
since 12,000 years BP. **Nature,** 292(5820):208-212.

Creer, K.M., Smith, G., Tucholka, P., Bonifay, E., Thouveny,
N. & Truze, E. 1986. A preliminary paleomagnetic study of
the Holocene and late Würmian sediments of Lac du Bouchet
(Haute Loire, France). **Geophys. Jour. Royal Astronom. Soc.
London,** 86 (1986):943-964.

Easterbrook, D.J. & Othberg, K. 1976. Paleomagnetism of
Pleistocene sediments in the Puget Lowland, Washington.
Inter. Geol. Corr. Progr., Rept. 3:198-207.

Ignatius, H. 1958. On the rate of sedimentation in the Baltic
Sea. **Extrait des Comptes Rendus de la Société Géologique de
Finlande** 30:135-144.

Kodama, K.P., Evenson, E.B., Clinch, J.M. & Rabassa, J. 1985.
Anomalous Geomagnetic field behaviour recorded by glacial
sediments from northwestern Patagonia, Argentina. **J. Geomag.
Geoelectr.,** 37:1035-1050.

Mercer, J.H. 1970. Variations of some Patagonian Glaciers since
the Late-Glacial: II. **Am. Jour. Scien.** 269:1-25.

Mercer: J.H. 1972. Chilean Glacial Chronology 20,000 to
11,000 Carbon-14 years ago: Some global comparisons. **Science**
176(4039):1118-1120.

Mercer, J.H. 1976. Glacial History of Southernmost South
America. **Quaternary Research** 6:125-166.

Mörner, N.A. 1982. Paleomagnetism of varved sediments of
Wisconsinan age from Ontario and Quebec, Canada. **Stockholm
Contributions in Geology,** XXXVII: 15, Stockholm 1982: 177-
199.

Mörner, N.A. & Sylwan, C.A. 1987. Revised terminal moraine
chronology at Lago Buenos Aires, Patagonia, Argentina.
IPPCCE Newsletter, 4:15-16.

Sinito, A.M., Valencio. D.A. & Creer, K.M. 1985. Paleolimnolo-
gía del area aledaña a los lagos El Trébol, Moreno y Nahuel
Huapi (Brazo Campanario), Provincia de Río Negro. **Asoc. Geol.
Arg., Revista,** XL(3-4):211-224.

CALVIN J.HEUSSER
New York University, USA

12

Quaternary vegetation of southern South America

ABSTRACT

Quaternary vegetational history of the higher latitudes of
Chile and Argentina (34°-56°S), as interpreted from fossil
pollen in stratigraphic sections, covers regions ranging from
present-day subtropical woodland and steppe to polar tundra.
Most sections are postglacial (Holocene) - late-glacial in age,
while some are from over 45,000 yr BP during the last
glaciation, and a few are from times of interglaciation. At
the time of the last glaciation, vegetation outside the glacial
border was comparatively open. Conforming to latitudinal and
altitudinal gradients, it consisted of tundra, steppe, parkland, and woodland. Tundra was extensive across Tierra del
Fuego and southern Patagonia for millennia of record prior
to 10,000 yr BP. To the north, evidence from temperate
latitudes of Chile indicates widespread parkland and tundra,
except during the middle of the last glaciation and in the
late-glacial when open woodland developed. On the east slope
of the subtropical Andes Mountains in Argentina, Patagonian
steppe extended farther north, while on the Chilean side,
woodland of southern beech and podocarp spread northward
across lowland which is today covered by broad-sclerophyllous
woodland and steppe. In the postglacial, closed forest
dominated by beech formed at higher latitudes, whereas at
lower latitudes, both in Argentina and Chile, reconstituted
scrub and grass steppe became a feature of much of the
landscape. Interglacial vegetation was characterized by open
woodland, at times rich in woody taxa, and by intervals of
impoverished grass steppe. There is no evidence for the
development of highly structured interglacial forest with

species complexity that is comparable to the modern rain
forest of southern coastal Chile.

RESUMEN

La historia de la vegetación cuaternaria de las latitudes más
altas de Chile y Argentina (34°-56°S), interpretada a partir
de polen fósil en secciones estratigráficas, cubre regiones
que varían del bosque tropical y la estepa hasta la tundra
polar. La mayoría de las secciones son postglaciales (Holoceno)
y tardiglaciales en edad, mientras que algunas son de más de
4500 años AP, durante la última glaciación y unas pocas son de
las épocas interglaciales. En el momento de la última glaciación, la vegetación que se encontraba por fuera del límite
externo glacial era comparativamente abierta. Conformándose
a los gradientes latitudinales y altitudinales, estaba inte-
grada por tundra, estepa, parque y bosque. La tundra fue ex-
tensiva a través de Tierra del Fuego y Patagonia meridional
por varios milenios de registro antes de 10.000 años AP. Hacia
el norte, evidencia procedente de las latitudes templadas de
Chile indica amplia estribación de parque y tundra, excepto
durante la mitad de la última glaciación y en el glacial tardío, cuando un bosque abierto se desarrolló. Sobre la pendien-
te oriental de los Andes subtropicales de Argentina, la estepa patagónica se extendió más al norte, mientras que en el
lado chileno, el bosque de **Nothofagus** y podocarpáceas se extendió al norte a través de las depresiones, las cuales están
hoy cubiertas por bosque de esclerófitos y estepa. En el postglacial, el bosque cerrado dominado por las fagáceas se formó
en las latitudes más altas mientras que en latitudes más bajas, tanto en Argentina como en Chile, la estepa herbácea y
el arbustal reconstituída se transformaron en el rasgo domi-
nante de la mayoría del paisaje. La vegetación interglacial
estaba caracterizada por bosque abierto, a veces rico en taxa
de ambientes boscosos y a intervalos, en pastos de la estepa
empobrecido. No hay evidencia del desarrollo del bosque interglacial fuertemente estructurado con una complejidad especí-
fica que es comparable a la moderna selva valdiviana del Chile litoral meridional.

INTRODUCTION

This review examines the literature on Quaternary palynology
at the southern end of the Americas and integrates attempts to
reconstruct the vegetation of the region. Geographic coverage
is poleward of 34°S, the parallel proximal to Santiago in
Chile and to Mendoza and Buenos Aires in Argentina, which forms
a natural boundary, north of which the density in number of
studies drops sharply. At 34°S until around 37°S in Chile
(sclerophyllous woodland) and to about 39°S in eastern Argentina (pampa or grass steppe), data are from subtropical sites.
Climate is semi-arid, except on the Atlantic side of the
continent in the transitional zone to winter rain (Miller, 1976;
Prohaska, 1976). Farther south, data are from temperate
localities. In Chile and the Argentine Andes (subantarctic
mixed southern beech forest), conditions are more humid,
whereas east of the Andes, they continue to be semi-arid
(Patagonian steppe). At the extreme end of South America in
southernmost Chile (moorland or tundra), which comes under the
influence of higher latitudes of the Southern Ocean, a few
sites are in places subject to polar climate (Pisano, 1981).

An overview is first presented of the evidence bearing on
Argentine vegetation from the stock of studies made on the
continent and on marine cores from the South Atlantic (Figure
1). Coverage for the Pacific slope of the Andes in Chile is
dealt with later in more expanded form (Figure 2). The review
emphasizes geographic distribution, extended time span, and
time control of pollen records, updating reviews by Markgraf
and Bradbury (1982) and Heusser (1983a). For morphological
treatment of modern pollen and spores of the region, reference
is made to the works by Auer **et al.** (1955), Heusser (1971),
Markgraf and D'Antoni (1978), and Wingenroth and Heusser
(1986), including citations therein.

ARGENTINA

1 FULL GLACIAL, LATE-GLACIAL AND POSTGLACIAL (HOLOCENE)

Diagrams for Tierra del Fuego reported by Von Post (1929,
1946) from sections collected in 1926-1927 by the Swedish
glacial geologist, Carl C:zon Caldenius, at Lago Fagnano,
Baño Nuevo, and Cabo Domingo (Figure 1), display the first

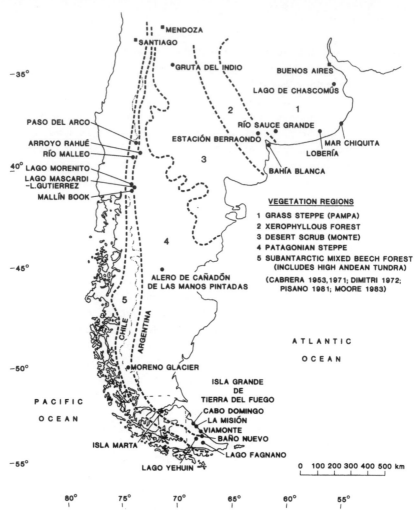

Figure 1. Southern South America showing locations of
Quaternary pollen sites discussed in the text in relation to
vegetation regions in Argentina.

fossil pollen data from South America (Caldenius, 1932). Their
interpretation stresses three-phase postglacial expansion,
dominance, and retreat of steppe vegetation, arguing for
regional parallelism of climatic changes on a global scale,
which was popular at the time. These seminal diagrams, today
of historical significance, set the stage for comprehensive
studies by Väinö Auer to follow.

200

Auer (1933, 1950, 1956, 1958, 1959, 1960, 1965, 1970, 1974), whose first field season was in 1928-1929, collected over 50 sections of late Quaternary mires in Patagonia (some of these in Chile) between 39°-54°S and 65 in Tierra del Fuego (a number of these also in Chile) between 53°-55°S. Their pollen records illustrate on a broad scale forest-steppe interaction along the Andean mountain front. On the assumption that three ash layers from regional volcanic eruptions found in the sections were chronostratigraphic, Auer outlined the maximum extent of forest from the proportion of tree pollen at about the time of eruption III, after which the boundary with the steppe receded. The tephrachronology is derived from radio-carbon ages of three volcanic ash layers in Tierra del Fuego at 9000 (eruption I), 5000 (eruption II), and 2200 yr. BP (eruption III).

The pattern set by Auer is recognizable in only certain of his data, and the concept he advanced of broadly uniform timing of eruptions is untenable because the tephrachronolo-gical base he used is not consistent. The number of ash horizons is so variable over the extensive region encompassed by the sections that only locally can ash serve for dating purposes. At La Misión (53°30'S), on the Atlantic side of the Fuegian steppe, Markgraf (1980a, 1980b) showed, for example, that ash assigned by Auer to the late-glacial proved to be early Holocene when its radiocarbon age was ascertained.

Because pollen records from the steppe are difficult to interpret, their value as sources of vegetation history is questionable. Surfaces of sites are subject to deflation and oxidation, thus rendering pollen in sediments incomplete and susceptible to differential preservation. Contrary to the observations made by Markgraf **et al.** (1981) and Mercer and Ager (1983), beech and other arboreal types are carried in quantities to the steppe by wind from Andean forest and wood-land. Their proportion can be significant, making up, for example, close to 25% of the pollen sum at sites along the eastern border of the Strait of Magellan, north of Tierra del Fuego, a distance of over 100 km from the forest border (C.J. Heusser, unpub. data). Fluctuations in percentages of tree pollen, therefore, are dependent not alone on pollen production, as related to the sizes of tree populations, but also on changes in wind strength and direction. For these reasons, interpretations of trends in pollen taxa at La Misión can be subject to error. Preference is for the record from Lago Yehuin (Markgraf, 1983), about 75 km south of La Misión, in the subantarctic deciduous forest (**Nothofagus pumilio, N. antarctica**) and steppe (**Festuca gracillima**).

Lago Yehuin (54°20'S), its chronology regulated by four radiocarbon dates, is an example of the three-phase Von Post-Auer model with some modification. Treeless grassland (8000-10,000 yr BP) at the beginning was followed by transitory spread of beech (**Nothofagus antarctica**) until around 6500 yr BP when an increase of grass suggests the maximum expanse of steppe. At 5000 yr BP, beech (including **N. pumilio**) began to increase again, so that after about 3000 yr BP, the proportion of trees along the forest-steppe boundary reached the modern level.

Northward in Argentine Patagonia, until Mallín Book at 41°20'S (Markgraf, 1983) is reached, dated postglacial records are few. A bog section in beech forest at Moreno Glacier (50° 30'S) is close to 9500 yr BP in age and shows extended beech dominance following an initial interval of herbs and shrubs (Mercer and Ager, 1983). At Alero del Cañadón de las Manos Pintadas (45°28'S) in the steppe, some 3000 yr of record, characterized by grass, chenopod, and composite, include increasing amounts of beech over the past 2500 yr (D'Antoni, 1978). Pollen in Auer's (1958) bog section in beech scrub between Lago Mascardi and Lago Gutierrez (41°15'S), dated at 8700 yr BP, also reveals a greater proportion of beech over late postglacial millenia (see data in Markgraf, 1983). Steppe elements in the section are conspicuous in the early post-glacial.

Deposition at Mallín Book, located in mixed beech (**N. dombeyi, N. pumilio**) forest and supplied with nine well-spaced radiocarbon dates, began following withdrawal of ice of the Nahuel Huapi glacier (Flint and Fidalgo, 1964). Earlier vegetation before 13,000 yr BP, in the main resembling steppe, consisted of mostly grass, composite, and heath (**Empetrum**). Expansion of beech (**N. dombeyi**) and cedar (**Austrocedrus type**), in evidence initially, occurred later, and by 12,000 yr BP, these and an occasional podocarp represented about 75% of the pollen sum (excluding sedge). Open woodland is suggested by the continuing presence of steppe until about 8500 yr BP. Closed beech forest (including **N. pumilio** and **N. antarctica**) is featured afterward until about 3000 yr BP, at which time, quantities of steppe elements increased. At Lago Morenito (41°03'S), 30 km to the north, the pollen sequence dated over 11,700 yr BP and resembling Mallín Book (Markgraf, 1984) indicates open beech-cedar forest over the last 5000 yr.

History of Araucaria (**A. araucana**) over the past 3000 yr
(Heusser **et al.**, 1986) derives from sections at Río Malleo
(39°36'S) and at Paso del Arco (38°52'S) within the geographic
range of the tree. Araucaria occurs in minor quantities (less
than 20%) associated with beech (**N. dombeyi** type) and grass.
Episodes of volcanism and fire have influenced distribution in
Araucaria in the past, as shown by its poor representation in
relation to layers of **tephra** and **charcoal** in the sections.
Recovery of Araucaria afterwards is indicative of its
adaptation to growth on soils of volcanic origin and its
resilience following burning.

Gruta del Indio (34°45'S), a rock shelter in desert scrub
(**Prosopis flexuosa, Larrea divaricata**) at the extreme north of
the region, contains an unusually long record with 16 radiocarbon dates over more than 32,000 yr (D'Antoni, 1980, 1983).
Principal component analysis of pollen in a 110 cm thick
deposit at the shelter identified grassland and scrub steppe
(Patagonian-type) before 10,000 yr BP. Desert scrub (Monte-
type), interrupted by riparian vegetation, was characteristic
afterward. Over the last 3000 yr at Gruta del Indio and at
other sites in the vicinity of Mendoza, Markgraf (1983) found
increasing proportions of Andean elements.

Remaining sections of note occur in the pampa (**Stipa
neesiana, Poa bonariensis, Piptochaetium montevidense**) at
Estación Berraondo (38°25'S, Quattrocchio **et al.**, 1983),
Lobería (38°10'S, Romero and Fernández, 1981), Bahía Blanca
(38°51'S, Guerstein and Quattrocchio, 1984), Laguna de Chascomús (35°40'S, Fernández and Romero, 1984), Mar Chiquita
(37°40'S, Nieto and D'Antoni, 1985), and Río Sauce Grande
(38°17'S, Rabassa **et al.**, 1986a). Data in most instances are
without time control or where dated by radiocarbon are of very
recent millennia. Some vegetational changes indicated by the
data appear to extend to the Pleistocene, as in the case of
Laguna de Chascomús, where inital halophytic communities were
succeeded by steppe and later pampa.

Support for a Pleistocene age of Laguna de Chascomús is
furnished by pollen in marine core RC12-241 from the Argentine
Basin in the South Atlantic Ocean (43°28'S, Heusser and
Wingenroth, 1984). Halophytic chenopod is abundant in sediments
estimated to date from 14,000-28,000 yr BP, including the time
of the last glacial maximum, and is followed by high
proportions of grass. Marine cores with extended records from
the Argentine Basin, reflecting conditions on the continent,

show chenopod and **Ephedra** during glacial stages, whereas
interglacial stages consist of much beech and podocarp (Groot
and Groot, 1966; Groot **et al.**, 1967; Stanley, 1967).

2 INTERGLACIAL

Auer (1933, 1956, 1970) assigned interglacial ages to three
peat deposits in Fuego-Patagonia. The deposits are exposed
at Viamonte (54°S) on the Atlantic coast of Isla Grande de
Tierra del Fuego, Isla Marta (53°S) in the Strait of Magellan,
and Arroyo Rahué (39°S) in northern Patagonia. Viamonte with
an age over 41,000 yr BP is a compressed, reworked peat,
resting between one underlying and two overlying tills. Its
age, stratigraphic position, and fossil content imply an
interglaciation with beech forest much in the same position
as today. Pollen of Araucaria and **Dacrydium** at Viamonte seems
certain to have been redeposited from Tertiary beds in which
both occur (Faseola, 1969). Isla Marta is also an intertill
deposit consisting of peat-bearing glaciofluvial sand. Undated
and conceivably of interglacial age, the peat contains pollen
of beech, grass, and sedge. At Arroyo Rahué, where fossil taxa
resemble modern regional taxa, reexamination of the site
establishes its age, dating between 27,000 and 33,000 yr BP,
as interstadial (Markgraf **et al.**, 1986).

CHILE

1 FULL GLACIAL, LATE-GLACIAL AND POSTGLACIAL (HOLOCENE)

Late-glacial tundra was extensive across the landscape of
much of Magallanes (Tierra del Fuego and southern Patagonia)
until about 10,000 yr BP (Heusser, 1984a). Data are from
two mires in subantarctic deciduous forest (Figure 2). One
at Puerto Williams (54°56'S), beginning at 13,000 yr BP and
encompassing some 3000 yr of tundra communities made up of
Empetrum (**E. rubrum**), grass, composite, **Acaena**, and umbellifer
(Figure 3), shows a large proportion of dwarf shrub tundra
evident at the onset, which is reduced late in the interval
by an increase in herb tundra. The other mire to the north at
Punta Arenas (53°10'S) shows ephemeral expansion of beech in
the tundra between 11,000-12,000 and before 13,500 yr BP. At

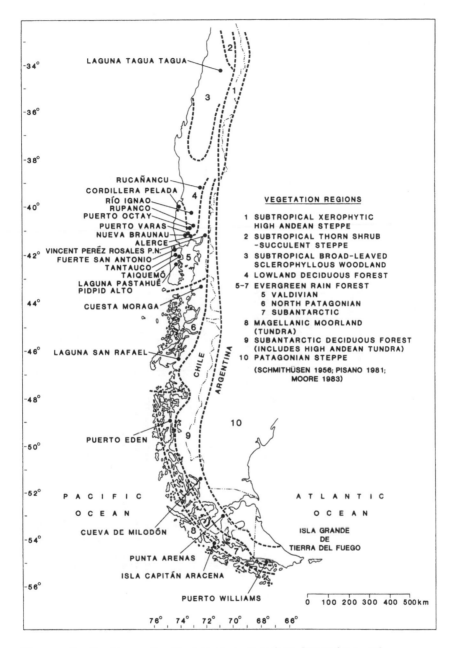

LAGUNA TAGUA TAGUA

RUCAÑANCU
CORDILLERA PELADA
RÍO IGNAO
RUPANCO
PUERTO OCTAY
PUERTO VARAS
NUEVA BRAUNAU
ALERCE
VINCENT PERÉZ ROSALES P.N.
FUERTE SAN ANTONIO
TANTAUCO
TAIQUEMÕ
LAGUNA PASTAHUÉ
PIDPID ALTO

CUESTA MORAGA

LAGUNA SAN RAFAEL

PUERTO EDEN

CHILE
ARGENTINA

PACIFIC
OCEAN

ATLANTIC
OCEAN

CUEVA DE MILODÕN

PUNTA ARENAS
ISLA CAPITÁN ARACENA

PUERTO WILLIAMS

ISLA GRANDE
DE
TIERRA DEL FUEGO

VEGETATION REGIONS

1 SUBTROPICAL XEROPHYTIC
   HIGH ANDEAN STEPPE
2 SUBTROPICAL THORN SHRUB
   -SUCCULENT STEPPE
3 SUBTROPICAL BROAD-LEAVED
   SCLEROPHYLLOUS WOODLAND
4 LOWLAND DECIDUOUS FOREST
5-7 EVERGREEN RAIN FOREST
   5 VALDIVIAN
   6 NORTH PATAGONIAN
   7 SUBANTARCTIC
8 MAGELLANIC MOORLAND
   (TUNDRA)
9 SUBANTARCTIC DECIDUOUS FOREST
   (INCLUDES HIGH ANDEAN TUNDRA)
10 PATAGONIAN STEPPE

   (SCHMITHÜSEN 1956; PISANO 1981;
    MOORE 1983)

0 100 200 300 400 500km

Figure 2. Southern South America showing locations of
Quaternary pollen sites discussed in the text in relation to
vegetation regions in Chile.

205

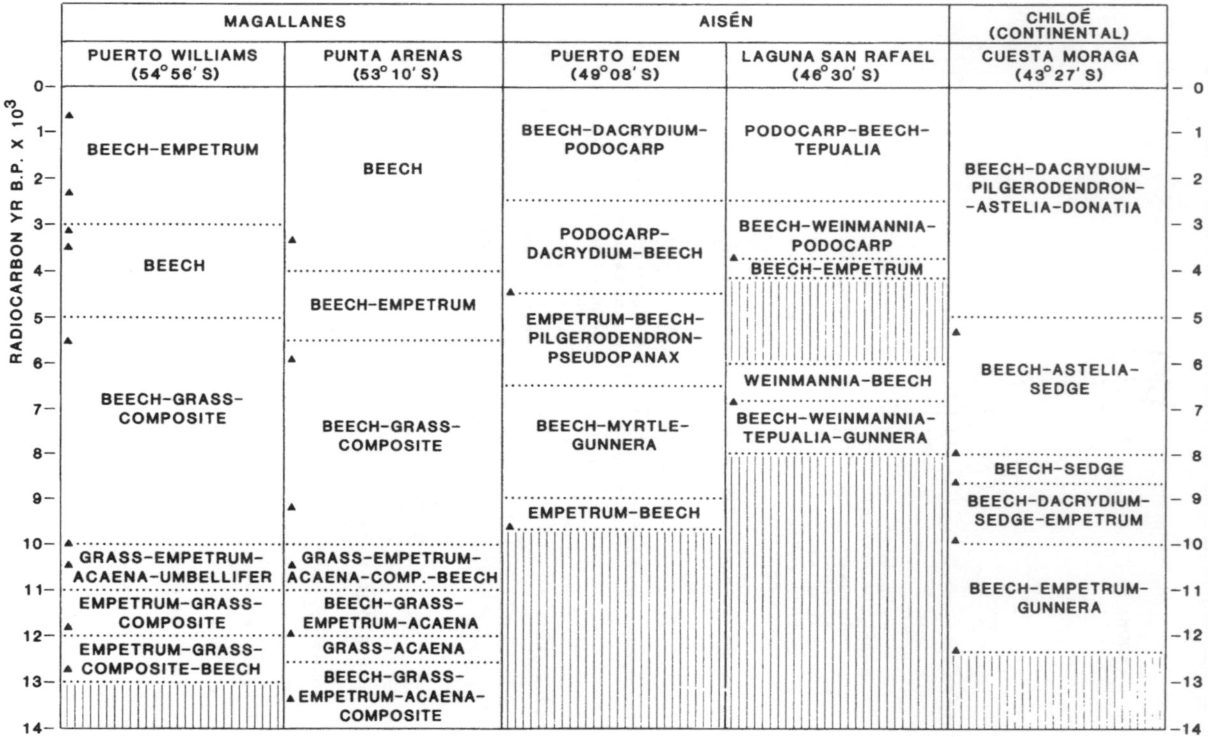

Figure 3. Correlation diagram of late Quaternary pollen assemblage zones in Magallanes, Aisén and Chiloé, Chile. Triangles indicate radiocarbon dated levels. See text for sources.

Isla Capitán Aracena (54°11'S), situated some 100 km south of Punta Arenas in Magellanic moorland (**Donatia fascicularis, Astelia pumila, Schoenus antarcticus**), beech expansion did not occur until later at about 9000 yr BP (Auer, 1974, original site designation is Isla Clarence).

During the early postglacial, when pollen influx was low at Puerto Williams (C.J. Heusser, unpub. data), an assemblage of beech-grass-composite is indicative of communities of relatively open woodland and steppe. Increase in beech influx after about 5000 yr BP at the time of the beech assemblage suggests that closed forest did not develop until the late postglacial. Hydrological and successional changes in the developing mire complex appear to account for assemblages of Empetrum with beech at Puerto Williams and Punta Arenas. Postglacial episodes of fire (Heusser, 1986) and higher sea level (Porter **et al.**, 1984; Rabassa **et al.**, 1986b) also influenced vegetation. The overall sequence since 12,400 yr BP is amplified by macrofossils studied by Moore (1978) at Cueva de Milodón (51°35'S).

Puerto Eden (49°08'S) in subantarctic rain forest (**Nothofagus betuloides, Pilgerodendron uviferum**) and Laguna San Rafael (46°30'S) in North Patagonian rain forest (**Weinmannia trichosperma, Laurelia philippiana, Podocarpus nubigena**) serve to trace the postglacial vegetational history of Aisén (Figure 3). Puerto Eden was at first represented by Empetrum and beech, after which beech-myrtle-Gunnera succeeded (Heusser, 1972a). Almost exclusive occurrence of beech during this interval may reflect greater extent of closed beech forest than now. Empetrum dominance later can be attributed to local edaphic control, possibly linked to climatic influence. Localities at lower latitudes are inferred as sources for late postglacial podocarp (**Podocarpus nubigena**) and Dacrydium (**D. fonkii**). At Laguna San Rafael in the late postglacial, podocarp gained prominence with beech and Tepualia (**T. stipularis**) following initial succession of beech and Empetrum on the end moraine of San Rafael Glacier (Heusser, 1960, 1964). After 6850 yr BP, communities of beech and Weinmannia were overridden when the glacier advanced during the Neoglaciation.

Cuesta Moraga (43°27'S) in Chiloé Continental (Figure 3) displays Dacrydium as early as 10,000 yr BP (C.J. Heusser, unpub. data). This mire, an enclave for Dacrydium, Astelia, and Donatia, is one of several stations in northern Patagonia Occidental where disjunctive populations of plants characteristic of Magellanic moorland are found. Cuesta Moraga

in the interior of the Andes was deglaciated before 12,300 yr
BP. Remarkably, no instance of fire is evident in the Cuesta
Moraga section, as charcoal is virtually absent. Multiple
tephra horizons record frequent volcanic eruptions.

The extented series of pollen assemblages at Taiquemó
(Figure 4) on Isla Chiloé (42°10'S), one of the longest
continuous records in Chile, dates from 43,000 yr BP (Heusser
and Flint, 1977). Taiquemó is a short distance beyond the
limit of the ice during the last glacial maximum (Llanquihue
Glaciation) at close to 20,000 yr BP (Mercer, 1984). Data at
this time imply a parkland of beech with tracts of tundra.
Earlier, before 31,000 yr BP, arboreal communities of beech-
podocarp-Pilgerodendron occupied Chiloé. At about 13,000 yr BP,
rain forest elements, particularly myrtle, began to arrive,
and by 9000 yr BP, components of Valdivian rain forest
(**Nothofagus dombeyi, Eucryphia cordifolia**) were established.
Villagrán (1985) has detailed vegetation about 60 km southwest of Taiquemó during late-glacial and postglacial time after
13,000 yr BP at Laguna Pastahue and also during glaciation at
around 36,000 yr BP at Pidpid Alto. Pollen from two undated
Chilotan sections (Godley and Moore, 1973), resulting from
collections made by the Royal Society Expedition to southern
Chile in 1958-1959, offers some additional source data
regarding postglacial vegetation.

The combined Puerto Varas-Alerce (41°19'-41°25') assemblages
of the past 16,000 yr in the "lake district' of Llanquihue
(Figure 4) are of beech parkland (13,000-16,000 yr BP) which
was later displaced by the expansion of woodland, consisting
of mostly myrtle, Weinmannia, beech, and Tepualia (Heusser,
1966, 1974). In the late-glacial - early postglacial, rain
forest species, both Valdivian and North Patagonian, including
Saxegothaea (**S. conspicua**), podocarp, Drimys (**D. winteri**),
Pseudopanax (**P. laetevirens**), and Eucryphia, continued
to invade. Rain forest, containing a major component of
Fitzroya (**F. cupressoides**), beech, Weinmannia, and podocarp,
appears to have had its optimal development during late postglacial millennia. The cordilleran counterpart of the
vegetational history over the past 9000 yr is contained in the
comprehensive account from Vincent Pérez Rosales Parque Nacional
in the Andes east of Puerto Varas-Alerce (Villagrán, 1980).

Farther north in the "lake district" of Osorno, 40-50 km from
Puerto Varas-Alerce, assemblages of three sections (Figure 4)
constitute a composite sequence (Heusser, 1974, 1981). These
are from an interstadial exposure (30,000-37,000 yr BP) and a

late-glacial - postglacial deposit in an abandoned spillway (dated 19,000 yr BP) at Puerto Octay (40°53'S) and from an interstadial exposure (20,000-24,000 yr BP) near Lago Rupanco (40°56'S). The older interstade is of beech parkland from which beech withdrew late in the interval prior to the deposition of Llanquihue drift. The younger interstade, preceding the last glacial maximum, is of tundra and beech parkland. Thereafter, as indicated by the spillway deposit assemblages, beech parkland spread over the western border of the lakes until about 12,500 yr BP. Myrtle-dominated woodland that followed is presumed to be of late-glacial age (possibly including the early part of the postglacial) as at Puerto Varas-Alerce. An unconformity in the upper part of the section (myrtle-beech-Weinmannia assemblage), separates near-surface and surface sediments which contain indicators of European settlement (**Pinus, Rumex, Lotus, Plantago**).

The Cordillera Pelada in Valdivia (40°10'S) with its relict populations of subantarctic species is regarded as a plant refuge during ice-age glaciation (Heusser, 1982). Donatia, typical of Magellanic moorland, is important in cushion bog assemblages which date from about 10,500 yr BP (Figure 4). Other fossil pollen representative of moorland consists of Dacrydium, Astelia, Gaimardia (**G. australis**), Drapetes (**D. muscosus**), and Drosera (**D. unifolia**). Drapetes did not persist in the Cordillera Pelada later than the early postglacial. Along with **Lycopodium fuegianum**, which was distributed in the southern "lake district" and on Isla Chiloé during the late-glacial (Heusser, 1972b), it is now limited in its range to southernmost South America. Cushion bogs of the Cordillera Pelada evidently expanded along with beech and Fitzroya in the late postglacial. Before this, stands of beech appear to have been more open and grassy.

Rucañancu (39°33'S) at the foot of the Andes in northeastern Valdivia and near the edge of the lowland deciduous forest (Heusser, 1984b) also exhibits considerable amounts of grass (Figure 4). Grass peaks at 8350 yr BP, followed by beech (**Nothofagus obliqua** type at first and later **N. dombeyi** type). Myrtle and Aextoxicon (**A. punctatum**) are important woodland components in the early postglacial. Podocarp (**Prumnopitys andina**), characteristic of the latter part of the late-glacial, today ranges in the Andes at the latitude of Rucañancu at elevations about 1000 m higher than it did 10,000-10,500 yr BP. This implies that forests on the west slope of the Andes where

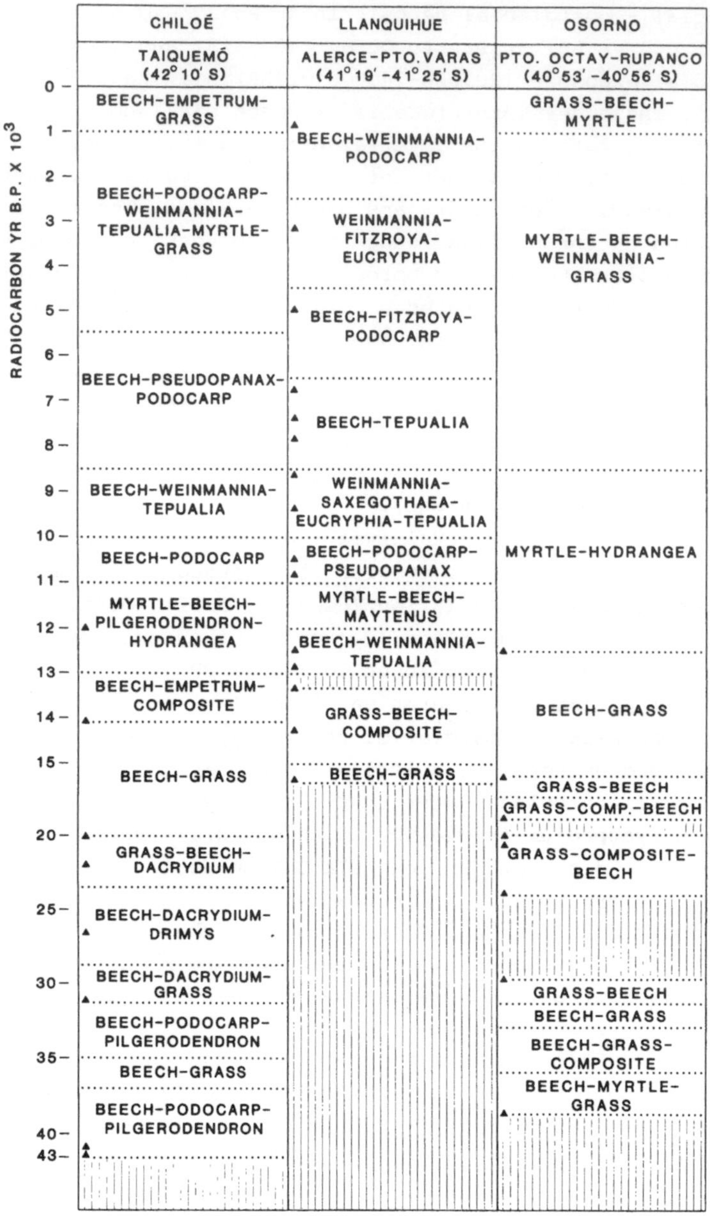

RADIOCARBON YR B.P. X 10³	CHILOÉ	LLANQUIHUE	OSORNO
	TAIQUEMÓ (42°10' S)	ALERCE-PTO.VARAS (41°19'-41°25' S)	PTO. OCTAY-RUPANCO (40°53'-40°56' S)
0–1	BEECH-EMPETRUM-GRASS		GRASS-BEECH-MYRTLE
1–2		BEECH-WEINMANNIA-PODOCARP	
2–5	BEECH-PODOCARP-WEINMANNIA-TEPUALIA-MYRTLE-GRASS	WEINMANNIA-FITZROYA-EUCRYPHIA	MYRTLE-BEECH-WEINMANNIA-GRASS
5–6		BEECH-FITZROYA-PODOCARP	
6–8	BEECH-PSEUDOPANAX-PODOCARP	BEECH-TEPUALIA	
9–10	BEECH-WEINMANNIA-TEPUALIA	WEINMANNIA-SAXEGOTHAEA-EUCRYPHIA-TEPUALIA	
10–11	BEECH-PODOCARP	BEECH-PODOCARP-PSEUDOPANAX	MYRTLE-HYDRANGEA
11–12	MYRTLE-BEECH-PILGERODENDRON-HYDRANGEA	MYRTLE-BEECH-MAYTENUS	
12–13		BEECH-WEINMANNIA-TEPUALIA	
13–14	BEECH-EMPETRUM-COMPOSITE	GRASS-BEECH-COMPOSITE	BEECH-GRASS
15	BEECH-GRASS	BEECH-GRASS	
16			GRASS-BEECH
17			GRASS-COMP.-BEECH
20	GRASS-BEECH-DACRYDIUM		GRASS-COMPOSITE-BEECH
25	BEECH-DACRYDIUM-DRIMYS		
30	BEECH-DACRYDIUM-GRASS		GRASS-BEECH
32	BEECH-PODOCARP-PILGERODENDRON		BEECH-GRASS
35	BEECH-GRASS		BEECH-GRASS-COMPOSITE
37			BEECH-MYRTLE-GRASS
40–43	BEECH-PODOCARP-PILGERODENDRON		

Figure 4. Correlation diagram of late Quaternary pollen assemblage zones in Chiloé, Llanquihue, Osorno, Valdivia and O'Higgings. Triangles indicate radiocarbon dated levels. See text for sources.

| VALDIVIA | | O'HIGGINS | RADIOCARBON YR B.P. X 10³ |
CORDILLERA PELADA (40° 10′ S)	RUCAÑANCU (39° 33′ S)	LAGUNA TAGUA TAGUA (34° 30′ S)	
BEECH–GRASS		CHENOPOD–GRASS–COMPOSITE	— 0
		GRASS–GUNNERA–UMBELLIFER–BEECH	— 1
DONATIA–BEECH–FITZROYA–MYRTLE	BEECH–GRASS	CHENOPOD	— 2 / — 3 / — 4
BEECH–FITZROYA–GRASS–DONATIA		GRASS–GUNNERA–UMBELLIFER–CHENO.	— 5
BEECH–GRASS–COMPOSITE	BEECH (N. OBLIQUA TYPE)–GRASS	GRASS–COMPOSITE–CHENOPOD	— 6 / — 7 / — 8
	GRASS–BEECH		— 9
	AEXTOXICON–BEECH	CHENOPOD–GRASS–COMPOSITE	
	MYRTLE–HYDRANGEA		—10
BEECH–DONATIA–GRASS	PODOCARP–MYRTLE		
	BEECH–GRASS–COMPOSITE		—11
		PODOCARP–BEECH–GRASS–CHENOPOD	—12 / —13 / —14
			—15
		BEECH–PODOCARP–GRASS–COMPOSITE	—20 / —25
		CHENOPOD–GRASS–COMPOSITE–BEECH	—30 / —35
		BEECH–PODOCARP–MAYTENUS–GRASS–COMPOSITE–CHENOPOD	—40 / —>43
		CHENOPOD–GRASS–COMPOSITE–BEECH	—>45

211

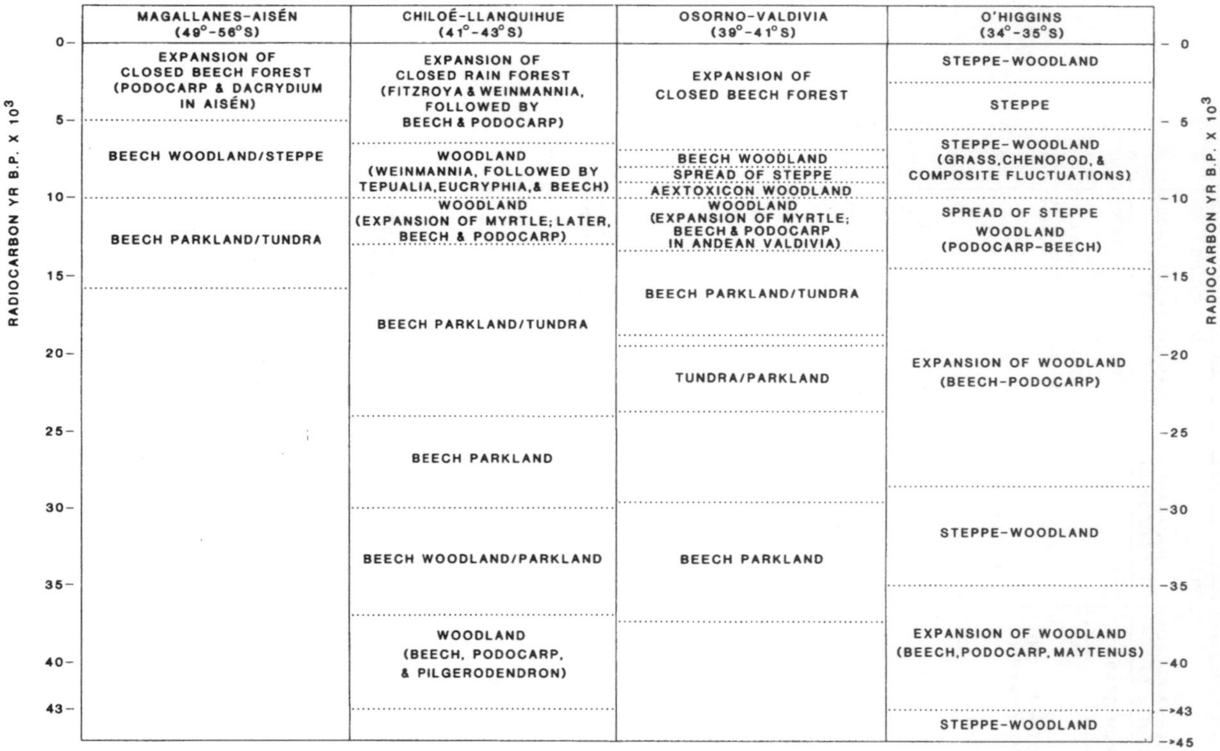

Figure 5. Late Quaternary vegetation sequences for provincial sectors of Chile.

RADIOCARBON YR B.P. X 10³

RADIOCARBON YR B.P. × 10³	MAGALLANES-AISÉN (49°-56°S)	CHILOÉ-LLANQUIHUE (41°-43°S)	OSORNO-VALDIVIA (39°-41°S)	O'HIGGINS (34°-35°S)
0	EXPANSION OF CLOSED BEECH FOREST (PODOCARP & DACRYDIUM IN AISÉN)	EXPANSION OF CLOSED RAIN FOREST (FITZROYA & WEINMANNIA, FOLLOWED BY BEECH & PODOCARP)	EXPANSION OF CLOSED BEECH FOREST	STEPPE-WOODLAND
5				STEPPE
	BEECH WOODLAND/STEPPE	WOODLAND (WEINMANNIA, FOLLOWED BY TEPUALIA, EUCRYPHIA, & BEECH)		STEPPE-WOODLAND (GRASS, CHENOPOD, & COMPOSITE FLUCTUATIONS)
10	BEECH PARKLAND/TUNDRA	WOODLAND (EXPANSION OF MYRTLE; LATER, BEECH, & PODOCARP)	BEECH WOODLAND / SPREAD OF STEPPE, AEXTOXICON WOODLAND / WOODLAND (EXPANSION OF MYRTLE; BEECH & PODOCARP IN ANDEAN VALDIVIA)	SPREAD OF STEPPE WOODLAND (PODOCARP-BEECH)
15		BEECH PARKLAND/TUNDRA	BEECH PARKLAND/TUNDRA	
20			TUNDRA/PARKLAND	EXPANSION OF WOODLAND (BEECH-PODOCARP)
25		BEECH PARKLAND		
30				STEPPE-WOODLAND
35		BEECH WOODLAND/PARKLAND	BEECH PARKLAND	
40		WOODLAND (BEECH, PODOCARP, & PILGERODENDRON)		EXPANSION OF WOODLAND (BEECH, PODOCARP, MAYTENUS)
>43 / >45				STEPPE-WOODLAND

212

podocarp grows at present have undergone a net upward shift
during postglacial time.

Laguna Tagua Tagua (34°30'S) in subtropical, broad-leaved,
sclerophyllous woodland of O'Higgings (Heusser, 1983b) is the
northernmost of the sections featured in this review. Dating
over 45,000 yr BP, probably from the middle of the last
glaciation, it is also the longest and most complete in Chile.
The record (Figure 4) shows significant percentages of arboreal
pollen (**Nothofagus dombeyi** type, **N. obliqua** type, **Prumnopitys
andina, Podocarpus saligna**) along with pollen of grass and
composite fluctuating with amounts of chenopod in the
Pleistocene, whereas during the postglacial, chenopod, grass,
and composite dominate, and arboreal types virtually disappear
over large segments. Intervals of arboreal dominance in the
Pleistocene between 14,500-28,500 and about 36,000-over
43,000 yr BP trace the spread of beech forest components from
the south and from the slopes of the Andes where they range
at present. The case of Prumnopitys, for example, at 33% of
the pollen sum 14,500 yr ago involved a range extension of at
least 180 km and a lowering of elevation of some 1000 m
(comparable to Rucañancu). Postglacial chenopod assemblages
at 8300-10,000, 2000-4500, and 0-1000 yr BP, alternating with
grass assemblages, imply vegetation more arid than now, when
levels of Laguna Tagua Tagua were low.

2 LATE QUATERNARY VEGETATION

As summarized by provincial sectors (Figure 5), considerable
openness distinguishes vegetation prior to 10,000 yr BP. From
Osorno-Valdivia south to Magallanes-Aisén, parkland and tundra
were of common occurrence during millennia centered on the last
glaciation. Tundra was distributed at higher latitudes and
in proximity to the cordillera and associated ice sheets and
glaciers. Woodland was increasingly in evidence before 30,000
yr BP in Chiloé-Llanquihue, developing optimally there between
37,000-43,000 yr BP and during the late-glacial in Chiloé-
Llanquihue and Osorno-Valdivia. In contrast to tundra and parkland in the south of Chile, vegetation of the central province
of O'Higgins was marked by interaction between steppe and
woodland. Significant is the fact that species of beech
(**Nothofagus dombeyi** type) and podocarp prospering in the
woodland no longer range so far equatorward or so low in
elevation. Vegetation today is mapped as broad-leaved

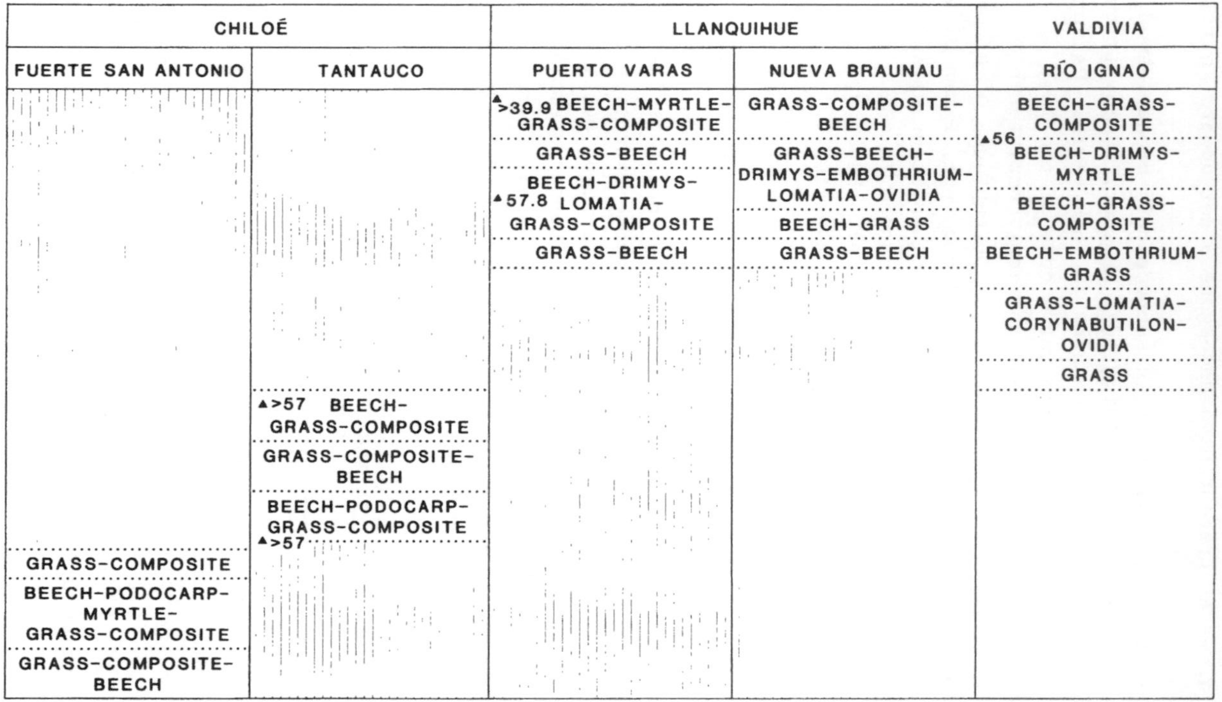

CHILOÉ		LLANQUIHUE		VALDIVIA
FUERTE SAN ANTONIO	TANTAUCO	PUERTO VARAS	NUEVA BRAUNAU	RÍO IGNAO
		▲>39.9 BEECH-MYRTLE-GRASS-COMPOSITE	GRASS-COMPOSITE-BEECH	BEECH-GRASS-COMPOSITE
		GRASS-BEECH	GRASS-BEECH-DRIMYS-EMBOTHRIUM-LOMATIA-OVIDIA	▲56 BEECH-DRIMYS-MYRTLE
		BEECH-DRIMYS-		BEECH-GRASS-COMPOSITE
		▲57.8 LOMATIA-GRASS-COMPOSITE	BEECH-GRASS	BEECH-EMBOTHRIUM-GRASS
		GRASS-BEECH	GRASS-BEECH	GRASS-LOMATIA-CORYNABUTILON-OVIDIA
	▲>57 BEECH-GRASS-COMPOSITE			GRASS
	GRASS-COMPOSITE-BEECH			
	BEECH-PODOCARP-GRASS-COMPOSITE ▲>57			
GRASS-COMPOSITE				
BEECH-PODOCARP-MYRTLE-				
GRASS-COMPOSITE				
GRASS-COMPOSITE-BEECH				

Figure 6. Correlation diagram of interglacial pollen assemblage zones in Chiloé, Llanquihue and Valdivia, Chile. Triangles indicate radiocarbon dated levels in millennia BP. See text for sources.

sclerophyllous at this subtropical latitude.

During the postglacial after 10,000 yr BP, open woodland in southern sectors was much supplanted by the expansion of closed forest. This replacement seems to have begun earlier in Osorno-Valdivia (around 7000 yr BP) than in Magallanes-Aisén (around 5000 yr BP). Beech has been consistently of importance in the forest, associated with podocarp and Dacrydium in Aisén and with Fitzroya, Weinmannia and podocarp in Chiloé-Llanquihue. To the north in O'Higgings, steppe was apparently widespread and varied in makeup, represented by grass, chenopod, and composite at different times. Evidence of sclerophyllous woodland is found both at the beginning and late in the post-glacial.

3 INTERGLACIAL

Section locations are in Chiloé, Llanquihue, and Valdivia (Figure 6). Two in Chiloé, Fuerte San Antonio (41°52'S) and Tantauco (42°08'S), are mapped outside the limit of the last (Llanquihue) glaciation (Heusser and Flint, 1977). At Fuerte San Antonio, sediments of apparent interglacial age interbedded in Fuerte San Antonio drift contain an arboreal component mostly of beech with small amounts of podocarp and myrtle. Equally important are grass and composite which are dominant early and late in the succession. At Tantauco, beech is throughout more abundant than grass and composite in a peat bed dated over 57,000 yr BP in intermediate glacial drift. Vegetation implied in both instances is of floristically depauperate, open woodland.

In Llanquihue (41°19'S), deposits at Nueva Braunau and Puerto Varas rest on pre-Llanquihue, Santa María drift (Porter, 1981). Nueva Braunau, beyond the Llanquihue drift border, is an eolian sequence laid down during and following wastage of the Santa María glacier. Asemblages are largely of grass in the lower part, followed by beech-grass, and of grass mixed with beech and a variety of arboreal species (notably **Drimys winteri, Embothrium coccineum**, and **Lomatia**) in the upper part (Heusser, 1981). Puerto Varas, an interdrift deposit from within the extent of the Llanquihue glacier, is dated over 39,900 yr BP; a finite age of 57,800 yr BP, apparently the result of sample contamination, is considered to be minimal. Sequentially, the species-poor assemblages principally of grass, at the beginning and later, are each followed by

assemblages of beech-grass with a comparatively rich
supplement of woody taxa. Both data sources imply intervals
of impoverished grass steppe and of open woodland dominated
by beech.

Data at Río Ignao (40°18'S) in Valdivia, assigned a minimal
age of 56,000 yr BP, are not unlike those at Nueva Braunau
and Puerto Varas. Grass, the almost exclusive taxon at the
start, is progressively displaced by woody plants, which first
consist of **Corynabutilon, Ovidia,** and **Lomatia** and later of
beech, supplemented by **Embothrium, Drimys,** and myrtle (Heusser,
1976). Grass and a wide assortment of arboreal and non-
arboreal taxa appearing with beech identify open mixed woodland that succeeded grass steppe. This pattern, replicable
in the data, seems to be characteristic of the interglacial
vegetation succession.

ACKNOWLEDGEMENTS

Supported by grants from the National Science Foundation to
New York University. Additional assistance was provided by
the Empresa Nacional del Petróleo and Servicio Nacional de
Geología y Minería in Chile and by the Argentine Consejo
Nacional de Investigaciones Científicas y Técnicas.

REFERENCES

Auer, V. 1933. Verschiebungen der Wald- und Steppengebiete
  Feuerlands in postglazialer Zeit. **Acta Geographica** 5:1-313.
Auer, V. 1950. Las capas volcánicas como base de la cronolo-
  gía postglacial de Fuegopatagonia. **Revista Investigaciones
  Agrícolas** 9:57-203.
Auer, V. 1956. The Pleistocene of Fuego-Patagonia. Part I.
  The ice and interglacial ages. **Annales Academiae Scientiarum
  Fennicae III. Geologica Geographica** 45:1-226.
Auer, V. 1958. The Pleistocene of Fuego-Patagonia. Part II.
  The history of the flora and vegetation. **Annales Academiae
  Scientiarum Fennicae III. Geologica Geographica** 50:1-239.
Auer, V. 1959. The Pleistocene of Fuego-Patagonia. Part III.
  Shoreline displacements. **Annales Academiae Scientiarum
  Fennicae III. Geologica Geographica** 60:1-247.
Auer, V. 1960. The Quaternary history of Fuego-Patagonia.
  Proc. Royal Soc. 152:507-516.

Auer, V. 1965. The Pleistocene of Fuego-Patagonia. Part IV. Bog profiles. **Annales Academiae Scientiarum Fennicae III. Geologica Geographica** 80:1-160.

Auer, V. 1970. The Pleistocene of Fuego-Patagonia. Part V. Quaternary problems of southern South America. **Annales Academiae Scientiarum Fennicae III. Geologica Geographica** 100:1-194.

Auer, V. 1974. The isorhythmicity subsequent to the Fuego-Patagonian and Fennoscandian ocean level transgressions and regressions of the latest glaciation. **Annales Academiae Scientiarum Fennicae III. Geologica Geographica** 115:1-88.

Auer, V., Salmi, M. & Salminen, K. 1955. Pollen and spore types of Fuego-Patagonia. **Annales Academiae Scientiarum Fennicae III. Geologica Geographica** 43:1-10.

Cabrera, A.L. 1953. Esquema fitogeográfico de la República Argentina. **Revista Museo la Ciudad Eva Perón** 8:87-168.

Cabrera, A.L. 1971. Fitogeografía de la República Argentina. **Boletín de la Sociedad Argentina de Botánica** 14:1-42.

Caldenius, C.C. 1932. Las glaciaciones cuaternarias en la Patagonia y Tierra del Fuego. **Geografiska Ann.** 14:1-164.

D'Antoni, H.L. 1978. Palinología del perfil del Alero del Cañadón de las Manos Pintadas. **Relaciones de la Sociedad Argentina de Antropología** 12:249-262.

D'Antoni, H.L. 1980. Los últimos 30 mil años en el sur de Mendoza, Argentina. **Memorias III Coloquio sobre Paleobotánica y Palinología. Instituto Nacional de Antropología e Historia,** p.83-108. México.

D'Antoni, H.L. 1983. Pollen analysis of Gruta del Indio. **Quaternary of South America & Antarctic Peninsula** 1:83-104.

Dimitri, M.J. 1972. Consideraciones sobre la determinaciór de la superficie y los límites naturales de la región andino-patagónica. In: M.J. Dimitri (ed.), **La región de los bosques andino-patagónicos,** p.59-79. Buenos Aires: Instituto Nacional de Tecnología Agropecuaria.

Faseola, A. 1969. Estudio palinológico de la Formación Loreto (Terciario Medio), provincia de Magallanes. **Ameghiniana** 6:3-49.

Fernández, C.A. & Romero, E.J. 1984. Palynology of Quaternary sediments of Lake Chascomús, northeastern Buenos Aires Province, Argentina. **Quaternary of South America & Antarctic Peninsula** 2:201-221.

Flint, R.F. & Fidalgo, F. 1964. Glacial geology of the east flank of the Argentine Andes between latitude 39°19'S and latitude 41°20'S. **Bull. Soc. Am.** 75:335-352.

Godley, E. & Moar, N.T. 1973. Vegetation and pollen analysis of two bogs on Chiloé. **New Zealand J. Botany** 11:255-268.

Groot, J.J. & Groot, C.R. 1966. Pollen spectra from deep-sea sediments as indicators of climatic changes in southern South America. **Marine Geology** 4:525-537.

Groot, J.J., Groot, C.R., Ewing, M., Burckle, L. & Conolly, J. R. 1967. Spores, pollen, diatoms, and provenance of the Argentine Basin sediments. In: **Progress in Oceanography,** 4:179-217. New York: Pergamon Press.

Guerstein, R. & Quattrocchio, M.E. 1984. Datos palinológicos de un perfil cuaternario, ubicado en el estuario de Bahía Blanca. **Actas Noveno Congreso Geológico Argentino** 4:595-609.

Heusser, C.J. 1960. Late Pleistocene environments of the Laguna de San Rafael area, Chile. **Geogr. Rev.** 50:555-577.

Heusser, C.J. 1964. Some pollen profiles from the Laguna de San Rafael area, Chile. In: L.M. Cranwell (ed.), **Ancient Pacific floras. The pollen story,** p. 95-114. Honolulu: Univ. Hawaii Press.

Heusser, C.J. 1966. Late Pleistocene pollen diagrams from the province of Llanquihue, southern Chile. **Am. Philos. Soc. Proc.** 110:269-305.

Heusser, C.J. 1971. **Pollen and spores of Chile.** Tucson: Univ. Arizona Press. 167pp.

Heusser, C.J. 1972a. An additional postglacial pollen diagram from Patagonia Occidental. **Pollen et Spores** 14:157-167.

Heusser, C.J. 1972b. On the occurrence of Lycopodium fuegianum during late-Pleistocene interstades in the Province of Osorno, Chile. **Bull. Torrey Bot. Club** 99:178-194.

Heusser, C.J. 1974. Vegetation and climate of the southern Chilean lake district during and since the last inter-glaciation. **Quat. Res.** 4:290-315.

Heusser, C.J. 1976. Palynology and depositional environment of the Río Ignao nonglacial deposit, Province of Valdivia, Chile. **Quat. Res.** 6:273-279.

Heusser, C.J. 1981. Palynology of the last interglacial-glacial cycle in mid-latitudes of southern Chile. **Quat. Res.** 16:293-321.

Heusser, C.J. 1982. Palynology of cushion bogs of the Cordillera Pelada, Province of Valdivia, Chile. **Quat. Res.** 17:71-92.

Heusser, C.J. 1983a. Quaternary palynology of Chile. **Quaternary of South America & Antarctic Peninsula** 1:5-22.

Heusser, C.J. 1983b. Quaternary pollen record from Laguna de Tagua Tagua, Chile. **Science** 219:1429-1432.

Heusser, C.J. 1984a. Late-Quaternary climate of Chile. In:
J.C. Vogel (ed.), **Late Cainozoic palaeoclimates of the
Southern Hemisphere,** p.59-83. Rotterdam: Balkema.

Heusser, C.J. 1984b. Late-glacial - Holocene climate of the
lake district of Chile. **Quat. Res.** 22:77-90.

Heusser, C.J. 1986. Fire history of Fuego-Patagonia. **Quaternary
of South America & Antarctic Peninsula.**

Heusser, C.J., Brandani, A., Rabassa, J. & Stuckenrath, R.
1986. Late-Holocene vegetation of the Andean Araucaria region,
Province of Neuquén, Argentina. **Mountain Res. & Development.**

Heusser, C.J. & Flint, R.F. 1977. Quaternary glaciation and
environment of northern Isla Chiloé, Chile. **Geology** 5:305-
308.

Heusser, L.E. & Wingenroth, M. 1984. Late Quaternary
continental environments of Argentina: evidence from pollen
analysis of the upper 2 meters of deep-sea core RC12-241
in the Argentine Basin. **Quaternary of South America &
Antarctic Peninsula** 2:79-81.

Markgraf, V. 1980a. New data on the late and postglacial
vegetational history of "La Misión", Tierra del Fuego,
Argentina. **Proc. IV Intern. Palynol. Conf.,** Lucknow, India
1976/1977, 3:68-74.

Markgraf, V. 1980b. Paleoclimatic reconstruction of the last
15,000 years in subantarctic and temperate regions of
Argentina. **Memoires Museum National Historia Naturelle,** B,
27:87-97.

Markgraf, V. 1983. Late and postglacial vegetational and
paleoclimatic changes in subantarctic, temperate, and arid
environments in Argentina. **Palynology** 7:43-70.

Markgraf, V. 1984. Late Pleistocene and Holocene vegetation
history of temperate Argentina: Lago Morenito, Bariloche.
Disertationes Botanicae (Festschrift Welten) 72:235-254.

Markgraf, V. & Bradbury, J.P. 1982. Holocene climatic history
of South America. **Striae** 16:40-45.

Markgraf, V., Bradbury, J.P. & Fernandez, J. 1986. Bajada de
Rahue, Province of Neuquén, Argentina: an interstadial
deposit in northern Patagonia. **Palaeogeogr., Palaeoclimatol.,
Palaeoecol.** 56:251-258.

Markgraf, V. & D'Antoni, H.L. 1978. **Pollen flora of Argentina.**
Tucson: Univ. Arizona Press.

Markgraf, V., D'Antoni, H.L. & Ager, T.A. 1981. Modern pollen
dispersal in Argentina. **Palynology** 5: 4363.

Mercer, J.H. 1984. Late Cainozoic glacial variations in South
America south of the equator. In: J.C. Vogel (ed.), **Late**

Cainozoic palaeoclimates of the Southern Hemisphere, p.45-68. Rotterdam: Balkema.

Mercer, J.H. & Ager, T.A. 1983. Glacial and floral changes in southern Argentina since 14,000 years ago. **Nat. Geogr. Soc. Res. Reports** 15:457-477.

Miller, A. 1976. The climate of Chile. In: W.Schwerdtfeger (ed.), **Climates of Central and South America. World Survey of Climatology 12**, p.113-145. Amsterdam: Elsevier.

Moore, D.M. 1978. Post-glacial vegetation in the south Patagonian territory of the giant ground sloth, Mylodon. **Bot. J. Linnean Soc.** 77:177-202.

Moore, D.M. 1983. **Flora of Tierra del Fuego.** Oswestry: Nelson. 396pp.

Nieto, M.A. & D'Anotni, H.L. 1985. Pollen analysis of sediments of the Atlantic shore of Mar Chiquita (Buenos Aires Province, Argentina). **Zbl. Geol. Palaont.** T.1, 1984. (11/12): 1731-1738.

Pisano, V.E. 1981. Bosquejo fitogeográfico de Fuego-Patagonia. **Anales del Instituto de la Patagonia** 12:159-171.

Porter, S.C. 1981. Pleistocene glaciation in the southern lake district of Chile. **Quat. Res.** 16:263-292.

Porter, S.C., Stuiver, M. & Heusser, C.J. 1984. Holocene sea-level changes along the Strait of Magellan and Beagle Channel, southernmost South America. **Quat. Res.** 22:59-67.

Von Post, L. 1929. Die Zeichenschrift der Pollen statistik. **Geol. Fören. Förhandl.** 51:543-565.

Von Post, L. 1946. The prospect for pollen analysis in the study of the earth's climatic history. **New Phytol.** 45:193-217.

Prohaska, F. 1976. The climate of Argentina, Paraguay, and Uruguay. In: W. Schwerdtfeger (ed.), **Climates of Central and South America. World Survey of Climatology 12**, p.13-112. Amsterdam: Elsevier.

Quattrocchio, M., Schillizzi, R. & Prieto, A. 1983. Quaternary sediments in the Estación Berraondo area (Buenos Aires Province, Argentina). **Quaternary of South America & Antarctic Peninsula** 1:105-112.

Rabassa, J., Heusser, C., Salemme, M., Politis, G. & Stuckenrath, R. 1986a. El hallazgo de troncos fósiles (Salix humboldtiana) en depósitos aluviales del Holoceno tardío , río Sauce Grande. **Ameghiniana,** Buenos Aires,in press.

Rabassa, J., Heusser, C. & Stuckenrath, R. 1986b. New data on Holocene sea transgression in the Beagle Channel, Tierra del Fuego. **Quaternary of South America & Antarctic Peninsula,** 4.

Romero, E.J. & Fernández, C.A. 1981. Palinología de paleosuelos del cuaternario de los alrededores de Lobería (provincia de Buenos Aires). **Ameghiniana** 18:273-285.

Schmithüsen, J. 1956. Die räumliche Ordnung der chilenischen Vegetation. **Bonner Geographische Abh.** 17:1-86.

Stanley, E.A. 1967. Palynology of six ocean-bottom cores from the southwestern Atlantic Ocean. **Rev. Palaeobot. Palynol.** 2:195-203.

Villagrán, M.C. 1980. Vegetationsgeschichtliche und pflanzensoziologische Untersuchungen in Vicente Pérez Rosales Nationalpark (Chile). **Dissertationes Botanicae** 54:1-165.

Villagrán, M.C. 1985. Análisis palinológico de los cambios vegetationales durante el Tardiglacial y Postglacial en Chiloé, Chile. **Revista Chilena de Historia Natural** 58:57-69.

Wingenroth, M. & Heusser, C. 1986. **Polen en la Alta Cordillera (Quebrada Benjamín Matienzo), Andes Centrales, Mendoza, Argentina.** Instituto Argentino de Nivología y Glaciología, Mendoza.

PAUL A.COLINVAUX
Ohio State University, Columbus, USA

13

Environmental history of the Amazon Basin

1 INTRODUCTION

Very little is known of the pattern of climate or vegetation in the Amazon Basin at times of maximum glaciation in
the northern hemisphere. From but a single station within
the entire basin has radiocarbon-dated paleoecological evidence of the last glaciation been produced, and this is
from the extreme western edge of the basin (Liu and Colinvaux, 1985). To this should be added radiocarbon-dated
sediments from the Atlantic Ocean and continental shelf
that yield data on minerals carried within the Amazon system, both in glacial times and since (Damuth and Fairbridge, 1970). Roughly 3500 km separate these sites and
we have no radiocarbon-dated evidence whatever for glacial
times from any station in between.

Despite this lack of data an hypothesis that the Amazon
lowlands were semi-arid in glacial times has come to prevail in contemporary debates about the Amazon past (Prance,
1982). This hypothesis rests on biogeographical arguments
put forward to explain disjunct distributions in the modern
Amazon biota, most particularly of birds (Haffer, 1974),
but is supported by a marshalling of various data on Amazonian soils or landforms that seem to require, at unknown
times in the past, local aridity beyond that attributable
to the dry periods of modern seasonal climates.

For the Holocene, or at least the later part of it, we
have more data, principally from the pioneering work of
Absy (1979, 1986), whose pollen studies at sites many
hundreds of kilometers apart provide a total of six radiocarbon-dated sections spanning up to 7000 B.P. Added to
these are six pollen and sedimentary histories from lakes
of the Ecuadorian Amazon, also spanning to 7000 B.P.,
(Frost, 1984; Colinvaux et al., 1985, and unpublished),
and radiocarbon-dated evidence for sporadic wildfires
provided by buried charcoal layers in the Venezuelan

Amazon (Sanford et al., 1984). These studies provide a
crude grid of sites yielding data that describe the magni-
tudes of Holocene climatic changes in the Amazon.

Clearly many more data are needed before reliable climatic
histories of the Amazon Basin can be constructed. The few
data yet available seem to suggest cooling, rather than aridity, as the general response of climate in the Amazon Basin
to northern glaciation, though present ecotonal climates
should have been marginally more arid. Millennial time-
scale climatic changes have been demonstrated for the Holocene and were almost certainly characteristic of Amazon
climates throughout the Quaternary.

2 CLIMATIC AND GEOGRAPHIC BOUNDARIES OF THE AMAZON BASIN

The Amazon Basin is a diverse land of continental size. Its
western flank is set by the twin, parallel cordilleras of
the Andes which form a continental and climatic divide
5-6000 m high. Mountainous regions of almost subcontinental
size border the northern and southern flanks:-- the Venezuelan and Guianan highlands, and the Matto Grosso. An immedi-
ate result of this configuration is that the Amazon Basin
is actually a set of several vast watersheds, each receiv-
ing run-off from its own unique mountain system. The principal watersheds (northern, southern, western, and their
subdivisions) drain the different climatic regions bordering on the Amazon Basin.

Although the Amazon river system drains all three flanks
of mountains into a single discharge at the Amazon mouth,
the main tributary systems are sufficiently distinctive to
be characterized by water chemistry and sediment loads
(Gibbs, 1967; Meade et al., 1979). Some of the local
peculiarities of watersheds and their climatic systems are
so noticeable as to be immediately evident, like the black
water, with its load of dissolved organic matter, in rivers
draining the Guiana Highlands and the Matto Grosso, or the
white-water streams, heavily charged with clay, silt, and
fine sand that drain the Andean slopes.

Precipitation can be high over large parts of the Amazon
Basin, reaching 5000-7000 mm per annum in the west. The
mechanism driving this high precipitation appears not to be
completely understood (Lamb, 1972). That the wettest areas
are associated with high relief suggests that orographic
factors are superimposed on the basic precipitation-inducing
mechanisms. The northern and southern watersheds have seasonal climates, with pronounced dry periods, particularly
to the east and the southwest. Rains do diminish in some
months of the year in the wetter regions of the west also
but a typical diminution of 40% in precipitation does not
provide a real dry season. In a region with 5 m of rain,

224

this still leaves precipitation an order of magnitude more
than effective evaporation (Landlivar, 1977).

The ultimate source of most Amazon precipitation is probably the Atlantic Ocean, as moisture-bearing air enters the
basin through its open, eastern flank. But Amazonian weather
is also under the influence of massive regional systems outside South America, like the southern oscillation, El Nino,
and movements of the intertropical convergence zone (ITCZ).
The possibilities for change in the impact of these systems,
both now and in glacial times, are strongly limited by the
permanent features of Amazon geography formed by the mountainous framework of the basin.

In the west, Amazon climates are partly insulated from
climates of the eastern Pacific by the Andes Mountains acting as a climatic divide 5-6000 m high. This barrier is
sufficient to deflect the climatic effects of El Nino years
from the western Amazon, preliminary data actually suggesting that precipitation may be out of phase on opposite sides
of the Andes, with coastal flooding of El Nino being associated with unusually dry years in the Peruvian Amazon (Thompson et al., 1984). Yet others of the effects of El Nino
and associate changes of the southern oscillation can be
traced across the Andes and northern South America in a broad
arc all the way to northeastern Brazil, particularly in vertical movements of the 200 millibar pressure line at about
12,000 m elevation, which can be correlated with relative
changes in precipitation (J. Rogers, Department of Geography,
The Ohio State University, personal communication of current
research results).

The whole Amazon Basin lies just to the south of the summer
position of the ITCZ which, where expressed, should cross
Venezuela and the Guiana Highlands. It is likely that precipitation in the northern Amazon is associated with annual,
southward, displacements of the ITCZ, with larger changes
in seasonal precipitation being probable as the ITCZ responds
to changes in the southern oscillation. The pattern of possible displacements of the ITCZ in South American longitudes
at times of glacial maxima is far from clear (Colinvaux,
1972; Newell, 1973). The possible implications for Amazon
climates are profound, however, as, for instance, a southward displacement by increased Hadley cell circulation should
increase precipitation in the northern Amazon.

3 TEMPERATURE DEPRESSION AT GLACIAL MAXIMA: THE COOLING
HYPOTHESIS

Global cooling is generally considered to be a characteristic
of glacial periods and assumptions of cooling underlie the
Melankovitch model. Cooling has been clearly demonstrated
for the Andes Mountains at elevations above 2000 m at about

5° N latitude where pollen data suggest a temperature depression of about 6° C in full glacial times (Van der Hammen and Gonzalez; 1960). Temperature depressions of this order in the high Andes are also suggested by depression of firn lines (Hastenrath, 1971) and by the budget for an extended Quelccaya ice cap (Thomp on et al., 1984). One of the longest pollen histories ever constructed suggests that in fact com parable temperature depressions in times of glacial advance have characterized the Colombian Andes in all glacial periods over the last 3.5 million years (Hooghimstra, 1984). Colombian tree-lines were raised to their present level, or above, in interglacial periods throughout the Quaternary.

The Andean records give credibility to the hypothesis that temperature depression was a property of glacial, equatorial South American climates at lower elevations also. A single data set from Mera, in Oriente Province of Ecuador, exists to examine this possibility. The data consist of fossil wood embedded in polliniferous sediments at 1100 m elevation, exposed in road cuts and radiocarbon-dated to 33,520 ± 1010 and 26,530 ± 270 B.P. (Liu and Colinvaux, 1985).

The Mera site is within the altitudinal range of lowland tropical rain forest in Amazonia, though approaching the upper limit (Grubb et al., 1963). Liu and Colinvaux (1985) used surface pollen samples to show that the modern rain forest near the Mera elevation of 1100 m is indeed comparable to that at elevations down to 250 m. Pollen from the fossil sections, however, differed markedly from these modern rain forest spectra. As with other pollen sections from the western Amazon lowlands, both fossil and modern pollen spectra from the Mera vicinity included numerous rare pollen taxa which could not be given botanical names. In addition, the pollen spectra are not directly comparable to any modern plant associations. But that the vegetation of Mera 33,000 to 26,000 B.P. differed markedly from modern Amazonian rain forest of the region is evident.

The Mera fossil pollen spectra have several properties in common with surface spectra from the Interandean Plateau in Ecuador at elevations above 2000 m, particularly in forest taxa. Podocarpus pollen is present and Alnus is prominent, with a near absence of the rain forest Cecropia. But the fossil spectra also differ from the Andean surface samples in that the spectra lack significant contributions from taxa like Gramineae, Chenopodiaceae and Amaranthaceae, or Rumex that apparently record the arid properties of modern Interandean vegetation, situated as it is in rain shadows from both cordilleras. Prominence of Cyperaceae in several Andean surface samples can be attributed to local overrepresentation from Scirpus totora beds surrounding the small lakes from which the samples were taken. When these arid or local properties of the modern Andean samples are set aside, a clear affinity of the fossil Mera spectra with Andean forest samples remains, strongly suggesting that the Mera site was occupied by a diverse, moist, montane forest thirty thousand years ago. A forest depression of at least several hundred meters is implied.

Wood samples included in the polliniferous deposits included softwoods thought to be of Andean species of Podocarpus (Liu and Colinvaux, 1985). It has since been pointed out that some species of Podocarpus exist in lowland forests, though none have been recorded from Ecuador (T. Stone and A.H. Gentry; personal communications). But these lowland podocarps appear to be constituents of lowland sites with pronounced dry seasons as in Rondonia of the southern edges of the Amazon Basin. The pollen assemblage from the Mera deposits, in which Podocarpus pollen appears in association with appropriate taxa of moist Andean forests, seems to confirm the original interpretation that the logs are of Andean Podocarpus species now prominent at elevations above 2000 m.

Since the lowest record for Podocarpus in Ecuador is at 1800 m, Liu and Colinvaux (1985) concluded that forest types were lowered by at least 700 m at 33,000 and 26,000 yr B.P., implying a minimal temperature depression of 4.5° C during times of glacial maxima. This temperature depression is minimal, both because the data give no reason for thinking that the Mera site was at the lowest elevation of the ancient Podocarpus stands and because the dating is consistent with the fossil forests having been of interstadial rather than full glacial age (op cit).

The Mera fossil wood and pollen provide the only radiocarbon-dated evidence for past climates yet available for any site within the Amazon Basin. The analysis from this one site, however, is consistent with the work in the high Andes in suggesting that temperature depression was a significant property of glacial climates at the equator, and allows extrapolation of the Andean results to low elevations.

These appear to be the only calculations of temperatures at glacial maxima to have been made for anywhere in wet equatorial lowlands, and the results conflict with the widespread belief that temperature depression at low latitudes and low elevations was modest. CLIMAP (1976) shows many areas of tropical oceans with surface temperatures within 2° C of modern temperatures at 18,000 yr B.P., letting many conclude that temperature depression was minimal at low elevation over equatorial land also. Close inspection of the CLIMAP results, however, does not support the contention that all equatorial ocean masses maintained near modern temperatures at glacial maxima. The true pattern is one of changed oceanic circulation in which water masses were displaced so that some equatorial water masses were colder than others. Furthermore, the actual data coverage of equatorial oceans is still sparse so that the map of ocean surface temperatures is far from complete. Temperature depression in the Amazon lowlands of between 4 and 6° C is not inconsistent with CLIMAP results (J. Imbrie, personal communication).

Thus the calculation of a minimal temperature depression of 4.5° C over the western Amazon during glacial maxima is fully in accord with what few circumstantial data exist and is in conflict with none. Very likely the true temperature depression in the lowland Amazonian forests of Ecuador at the last glacial maximum was in the order of 6° C as in the Colombian Andes.

Models of equatorial climates of a cooled earth have been made by applying Milankovitch forcing to the General Circulation Model (GCM) (Kutzbach and Guetter, 1984 and 1986). No conclusions are available for the generality of Amazonian climates, particularly of the wetter regions, but a general result is that seasonal, monsoon rains decrease by about 10% from modern values. This suggests that peripheral areas of Amazonia, with strongly seasonal climates in modern times, may have experienced somewhat more pronounced dry seasons at times of glacial maxima, but the maximal reduction of monsoonal rains possible from this effect would be in the order of 20% (J. Kutzbach, personal communication).

Based on the scanty, available, radiocarbon-dated evidence, therefore, the prime property of Amazon climates in times of northern glaciation was that they were cooler than Holocene climates by up to 6° C. This suggests that the wettest uplands of modern Amazonia, like those postulated to have been rain forest refugia by Haffer (1974), may well have supported moist forests, but that these should have differed from tropical rain forests so markedly that they were unlikely to have offered habitats to many species of the modern tropical rain forests.

4 THE HYPOTHESIS OF EQUATORIAL ARIDITY

Contemporary discussions about the ice-age Amazon have revolved about the likelihood of there having been a semiarid climate in the Amazon lowlands that replaced modern tropical rain forest with seasonal woodland, or even savanna, at times of glacial maxima (Simpson and Haffer, 1978; Colinvaux, 1979; Prance, 1982). This concept of a semiarid Amazon rests very largely on an hypothesis of refugia for rain forest taxa in postulated wet enclaves, as put forward by Haffer (1969, 1974) to explain disjunct distributions of modern taxa. This is discussed briefly in the next section. But the hypothesis of equatorial aridty rests also on observations for aridity in glacial times elsewhere in the tropics, and on landform data suggesting local aridity within parts of the Amazon Basin at unknown times in the past.

Dry climates during glaciations in regions the climates of which are now moist have been demonstrated in various tropical sites, most notably in Africa where some modern habitats of lowland rain forest held dry woodlands or savannas in the last glacial maximum (Livingstone, 1975). Similar reports from Central America or Asia would allow putting together a portfolio of tropical sites known to have been drier in glacial times.

From tropical South America come data for ice-age aridity in climate systems both west and north of the Amazon Basin. That from the Galapagos Islands to the west is the strongest, since radiocarbon control shows that a climate more arid than that of the Holocene endured on the Galapagos from at least the time of the mid-Wisconsin interstade until 10,000 B.P. (Colinvaux, 1972), and our unpublished pollen data from sediments of a lake within the Ecuadorian Interandean Plateau now suggest a modest increase in aridity on the Pacific coast of South America in synchrony with that of the Galapagos. The local mechanism causing this Galapagos and eastern Pacific aridity is still not known with certainty and proffered explanations include both northward and southward displacements of the ITCZ or changes in oceanic circulation (Colinvaux, 1972; Newall, 1973). None of these mechanisms require aridity in the Amazon Basin across the other side of the Andean divide.

From north of the Amazon Basin late-glacial aridity is convincingly demonstrated at Lake Valencia in the Venezuelan highlands (Bradbury et al., 1981; Salgado-Labourieu, 1980; Leyden, 1985). The basin of modern Lake Valencia was essentially swamp about 13,000 B.P., the surrounding vegetation being dry woodland or savanna. The present deep lake formed in the early Holocene and vegetation changed to tropical moist forest. The region now has a markedly seasonal climate. A pollen history from coastal British Guiana suggests drier vegetation in late glacial times in what is now mangrove swamp (Wijmstra and van der Hammen, 1966). Both histories are from a region to the north of the ITCZ, presumably influenced by Caribbean weather patterns, and separated from the Amazon Basin by high relief of subcontinental proportions.

Just as one can marshall data illustrating tropical sites that were arid during northern glaciations, so too can sites with wetter climates be listed and bordering on the Amazon Basin, the obvious data come from the Andes, where the Sabana de Bogota was flooded during classical Wisconsin times (van der Hammen and Gonzalez, 1960). This does not mean that the Amazon lowlands were extra wet in glaciations; rather it shows that regions separated by relief, or in adjoining climatic systems, should not be expected to have climates that change to wet or dry in synchrony. There are no general reasons for expecting equatorial aridity to be an ubiquitous property of the ice-age earth, except in monsoonal areas where, as described above, precipitation was expected to be reduced by about 10% from modern values. Otherwise we should

expect local or regional climatic shifts that may reduce precipitation in some places and increase it in others.

Yet there are empirical data from the Amazon Basin itself
that suggest past local aridity. The most interesting, because dated to glacial times, is the record from submarine
cores taken from the deep sea and continental shelf in the
Atlantic Ocean off the mouth of the Amazon River by Damuth
and Fairbridge (1970). Sediments of Amazonian origin reaching the sea during the last glacial maximum can be identified with certainty. The mineral composition of glacial-
age sediments differs from that of Holocene sediments in that
they include a fraction of minerals strongly suggestive of
erosion of an arid land surface. This fraction of the sediments, of course, could have come from any of the several
different watersheds draining into the Amazon delta, and
Damuth and Fairbridge concluded, after thorough review, that
the arid land component of the offshore sediments were products of erosion from the Guiana Highlands, not from lowland
Amazonia. This is entirely consistent with what we know of
the modern Amazon system where the northern drainage is
through dry terrain and where the pollen record from Lake
Galheiro shows persistent savanna in the late Holocene (Absy,
1979). If this ecotonal region was subject to the 10% diminution in precipitation modelled by Kutzbach and Guetter
(op cit.), then the conclusions of Damuth and Fairbridge are
convincing. Thus the record of offshore sediments is not
evidence of aridity in the Amazon lowlands but rather suggests more severe dry seasons along the ecotonal boundary of
northern Amazonia.

A single pollen history from Amazonia has been cited as
suggesting savannas in the Amazon lowlands. This is the diagram from old sediments in Rondonia towards the extreme
south west part of the basin obtained by Absy and van der
Hammen, (1976). Local habitats support tropical moist forest at present but pollen from the sediments shows that
savanna, or at least more open vegetation, occupied the site
at some unknown time in the past. Unfortunately the section
is without radiocarbon dates and there is no more reason to
think that the savanna episode is of the glacial maximum
than there is that it is part of a Holocene sequence. More
importantly, the region now has a strongly seasonal climate
and modern maps made by the Brazilian side-scanning radar
project (Projeto Radambrazil, 1978) show patches of savanna
surrounded by tropical forest within 100 km of the site. In
this ecotonal region the aridity mechanism of Milankovitch
forcing might alter the local pattern of savanna and forest
to produce the observed pollen history. It is equally likely
that the record describes ecotonal changes consequent on
significant climatic events during the Holocene.

There remain intriguing reports of stone pavements and
fossil sand dunes from various parts of the Amazon lowlands

230

(Ab'Saber, 1982; Bigarella and Andrade-Lima, 1982). Apparently none of these finds are in contexts that let them be dated and there is nothing to show that the land forms date to any particular part of the late-Quaternary, whether glacial or Holocene. It will be important to know how their distribution relates to the patterns of modern dry seasons in the several Amazon watersheds.

In the absence of radiocarbon-dated evidence, the circumstantial case for widespread, lowland aridity in Amazonia during glacial maxima is less than convincing. Were it not for the attraction of the refugial hypothesis put forward by Haffer (1974), these circumstantial data could scarcely have led to the aridity hypothesis.

5 THE REFUGIAL HYPOTHESIS

The refugial hypothesis was originally put forward by Haffer (1969, 1974) to account for disjunct bird distributions, but has since been extended to use the data of Anolis lizards, Helioconus butterflies, and various other taxa, both animal and plant (Simpson and Haffer, 1978; Prance, 1982). The disjunct distributions of different groups of taxa are not exactly synchronous or even concentric, but they do overlap, and a common denominator in the disjunct distributions of quite unrelated taxa is that they seem to reflect the same underlying physical relief. In particular, each of the principal areas of disjunction appears to be centered on a region of moist climate associated with relative high elevation. Haffer (op. cit.) postulated that these wetter regions remained moist and warm throughout glacial cycles, thus providing permanent habitats for rain forest species. Because the distributions were indeed disjunct, it followed that the intervening regions should be less permanent, or satisfactory, for rain forest species so that there should be no exchange of species between refugia. Haffer and his school argue that migration barriers between regions of disjunction have not been adequate in the recent past for the disjunctions to develop and they postulate that climatic changes of glacial cycles erected suitable barriers with aridity in the Amazon lowlands (Prance, 1982).

So marked is the influence of relief that many of the disjunctions, particularly of butterfly taxa, are actually of distributions completely outside the Amazon Basin, reflecting species populations in the high Andes, the Matto Grosso, or the Guianan highlands. That species disjunctions should reflect landforms and climate seems logical. Granted that the different Amazon watersheds have great physical and climatic diversity, it may be that an explanation for the species disjunctions can be found in contemporary geography and ecological processes. For Quaternary scientists the

231

refugial hypothesis will have widespread interest only to the extent that it can be tested with independently dated evidence for past climates. The only test so far made, that of the Mera record within the territory of the proposed Napo refugium of Haffer (1974), shows that the tropical rain forest history there was interrupted in the last glacial cycle by an episode of cool montane forest (Liu and Colinvaux, 1985).

6 FLOODING IN THE WESTERN AMAZON, 1300-800 B.P.

Lake sediment records from close to the equator in Ecuador have been taken to show regional flooding in the western Amazon 1300-800 yr B.P. (Colinvaux et al., 1985). The four lakes studied occupy deep sections of old channels abandoned by the neighboring rivers. Two of the lakes lie close to the Andean foothills, although well down in the lowland rain forest at 330 m elevation, and the other two are about 100 km further away from the foothills to the east at about 230 m elevation. Sediment cores show that all four lakes have deposited organic-rich gyttja continuously for the last 7-800 radiocarbon years. Underlying the gyttja in all four lakes is grey mineral sediment comparable to sediment in the modern river beds, showing that all four channels were occupied by the parent rivers before 7-800 B.P. In one of the lakes, old riverine sediment is itself underlain by peaty gyttja, the boundary having a radiocarbon age of about 1300 B.P. Colinvaux et al. (1985) concluded that there had been a regional flooding event in the western Amazon resulting from increased precipitation throughout the 500 yr of the interval from about 1300-800 B.P.

An event of the magnitude of the proposed 1300-800 B.P. flooding should be detectable over a wide area. Pollen data from endorheic lakes of the adjacent cordilerra of the Andes now strongly suggests heavy precipitation on the eastern flank of the Andes in this radiocarbon-dated interval (Colinvaux et al., unpublished results). But perhaps more interesting is detection of the flooding event downstream in the central Amazon region where Absy (1979) has two of her sections, Costa da Terra Nova and Lake Caju.

The 2800 yr Costa da Terra Nova record is from a 19 m section of river channel through a varzea which shows an abrupt change at the 10 m level from "fine clayey sand" to "grey clay with yellow veins". Extrapolation from the single radiocarbon-date dates this change to about 1400 B.P., about the time of the onset of flooding in the Ecuadorian lakes. Faster flow in a deeper channel would account for the change in sediment type.

The Lake Caju record is also from a varzea, but this time from a more closed basin. A trench in the broad sandy beaches

232

of the varzea revealed that clay-rich sediment was sandwiched
between a few cm of surface sand and a thick sand layer at
the base of the trench. A distinct pollen zone within the
clay-rich layer has striking peaks of Myrtaceae and Miconia
pollen in a pollen diagram that otherwise appears to represent vegetation of the sandy flats and surrounding varzea
woodland. Absy (1979) interpreted this Myrtaceae-Miconia
zone as a wetter period when rain forest elements invaded
the varzea or lake margins. Her interpolated dates for this
wetter episode are 1500-800 B.P.; essentially the same interval assigned to the western Amazon flooding event.

Costa da Terra Nova and Lake Caju are the only two of
Absy's sections to lie in the path of drainage from the Andean
watershed. Both have evidence of having received flood waters
in the 1300-800 B.P. interval but none of her sections in
drainages from other watersheds do so. Thus the effects of
high precipitation in the Andes, and flooding in the western
Amazon, can be detected downstream as predicted.

At least one of the climatic systems impinging on the Amazon
Basin, therefore, has been shown to be capable of wide fluctuation within the Holocene. Doubtless comparable changes,
both of increased precipitation and of reduced precipitation
have resulted from changes in climate at the western boundary,
both during the Holocene and before.

7 HOLOCENE ARID EPISODES AND POLLEN RECORDS

The most dramatic demonstration that unusual dry periods can
occur under Holocene conditions in Amazon is the recent demonstration by Sanford et al. (1985) that lowland rainforest is
subject to wildfire. Apparently a period of twenty or more
days with no rain at all is required, in a climatic regime
that either has a negligible dry season or within the normally wet season of a climate that does have a dry season.
Trees then can no longer transpire; they wilt, and drop their
leaves, which then become a layer on top of the root mat
subject to spontaneous combustion (R. Sanford, personal communication). By dating charcoal horizons in soils, Sanford
et al. (1985) demonstrated the intermittant occurrence of
wild fires at 250, 350-400, 640, 1500, 3000, and 6200 B.P.
The earliest known human occupation of the region is Sanford
et al.'s (1985) own dating of a pot sherd by thermolumines-
cence to 3750, strongly suggesting that the 6200 B.P. fire
antedated human arrival and thus giving credence to the
postulate that the other fires in apparently virgin forest
soils were not set by humans either.

Episodes of increased aridity during the Holocene are
described by Absy (1979, 1986) from her remarkable study of
six, radiocarbon-dated sections of parts of the Holocene
from widely separated parts of the Amazon Basin. Absy has

sites in the partly arid, southward flowing drainage from the
Guianan highlands (Lakes Galheiro and Cumina), from the
strongly seasonal southwestern drainage (Lake Surara), in the
southern drainage from the Matto Grosso region (Lake Arari),
and in the path of the main drainage of the western Amazon
region itself (Costa da Terra Nova and Lake Caju. The lakes
sampled are all varzeas or otherwise in the paths of seasonal
flooding and thus are particularly well suited to record
changes in water coming down the rivers with the annual wet
periods of the more seasonal watersheds. Drier episodes are
recorded in pollen diagrams in various ways; as vegetation of
bare flats, as stranded floating mats, or as eqpisodes of
colonization by Cecropia. Increased aridity is recorded at
all her sites at times past, mostly before 2000 B.P.

With considerable changes of sedimentary regime in many of
the varzea sections it is difficult to be precise about dating of particular episodes of dryness, since only a single
radiocarbon date is available from most of Absy's sections.
Because the varzeas are in drainages of the several major
watersheds of the Amazon system it is perhaps unlikely that
arid episodes will prove to be synchronous at the various
sites. What has been demonstrated is that seasonality can
fluctuate widely throughout most of the watersheds of the
Amazon Basin that were subjected to seasonal climates in th
Holocene. Coupled with the demonstration of past fires by
Sanford et al. (op. cit.), these pollen sections demonstrate
the variability of Holocene climates throughout the Amazon
Basin.

8 CONCLUSION: THE LITTLE THAT WE KNOW ABOUT THE CLIMATIC
HISTORY OF THE AMAZON BASIN

The data are still too few for any but the most tentative
suggestions about Amazon climates before about a few thousand
years ago. The principal climatic change of glacial maxima
was probably cooling that replaced tropical rain forest on
wetter uplands of the basin with forest associations appropriate to wet habitats with mean temperatures as much as 6°
colder. Increased aridity was probably significant only in
modern ecotonal areas that have pronounced dry seasons today.
Other parts of lowland Amazonia may have supported wet tropical rain forest then as now. The few dated paleoecological
data so far obtained make it doubtful that rain forest taxa
were reduced to wet refugia in highlands separated by arid
lowlands.

Records from the latter half of the Holocene are more in-
formative, suggesting that millennial time-scale changes in
precipitation in the various Amazonian watersheds have
occurred and that they can be both prolonged and of wide
amplitude. It is possible that some of the surface soil

phenomena attributable to dry episodes record some of the drier of these local Holocene episodes.

There is an urgent need for more, long, radiocarbon-dated records from all parts of the Amazon Basin. As these are planned, it will be important to investigate each of the several vast watersheds of the Amazon Basin, because most of these impinge on the different climatic systems that bound the Amazon Basin. Until these records are produced it is necessary to be cautious about interpreting past Amazon climates.

9 ACKNOWLEDGMENTS

My work on Amazon paleoecology has developed in close association with M. Steinitz-Kannan, Kam-biu Liu, I. Frost, and M. Miller. H. and T. Steinitz of Quito provided the base from which the Ecuadorian work of my laboratory was based. The country and people of Ecuador provided hospitality in numerous ways, and I give thanks to that pleasant land.

REFERENCES

Ab'Saber, A.N. 1982. The paleoclimate and paleoecology of Brazilian Amazonia. In G. Prance (ed.), Biological diversification in the tropics, p. 41-59. New York: Columbia.

Absy, M.L. 1979. A palynological study of Holocene sediments in the Amazon Basin. Amsterdam.

Absy, M.L. 1986. Palynology of Amazonia: the history of the forests as revealed by the palynological record. In G. Prance & T. Lovejoy (eds.), Amazonia, p. 72-82. Oxford: Pergamon.

Absy, M.L., and van der Hammen, T. 1976. Some paleoecological data from Rondonia, southern part of the Amazon Basin. Acta Amazon. 6: 293-299.

Bigarella, J.J. & D. de Andrade-Lima 1982. Paleoenvironmental changes in Brazil. In G. Prance (ed.), Biological diversification in the tropics. New York: Columbia.

Bradbury, J.P., B. Leyden, M.L. Salgado-Labouriau, W.M. Lewis, C. Schubert, M.W. Binford, D.G. Frey, D.R. Whitehead & F.H. Weibezahn 1981. Late Quaternary environmental history of Lake Valencia, Venezuela. Science 214: 1299-1305.

CLIMAP Project Members 1976. The surface of the ice-age Earth. Science 191: 1131-1137.

Colinvaux, P.A. 1972. Climate and the Galapagos Islands. Nature, 240: 17-20.

Colinvaux, P.A. 1979. Ice Age Amazon. Nature 278: 399-400.

Colinvaux, P.A., M.C. Miller, K-b. Liu, M. Steinitz-Kannan & I. Frost 1985. Discovery of permanent Amazon lakes and hydraulic disturbance in the upper Amazon Basin. Nature, 313: 42-45.

Damuth, J.E. & R.W. Fairbridge 1970. Equatorial Atlantic deep-sea arkosic sands and ice-age aridity in tropical South America. Bull. Geol. Soc. Am. 81: 189-206.

Frost, I. 1984. A paleolimnological and palynological investigation in the Ecuadorian rain forest: evidence of regional flooding and paleohydrological disturbance in the Amazon rain forest. Thesis, Columbus: Ohio State Univ.

Gibbs, R.J. 1967. The geochemistry of the Amazon river system: part I. The factors that control the salinity and the composition and concentration of the suspended solids. Geol. Soc. Am. Bull. 78: 1203-1232.

Grubb, P.J., J.R. Lloyd, T.D. Pennington & T.C. Whitmore 1963. A comparison of montane and lowland forests in Ecuador. 1: the forest structure, physiognomy, and floristics. J. Ecol. 51: 576-601.

Haffer, J. 1969. Speciation in Amazon forest birds. Science 165: 131-137.

Haffer, J. 1974. Avian speciation in tropical South America. Cambridge: Nuttal.

Hastenrath, S. 1971. On the Pleistocene snowline depression in the arid regions of the South American Andes. J. Glaciol. 10: 225-267.

Hooghiemstra, H. 1984. Vegetational and climatic history of the high plain of Bogota, Columbia: a continuous record of the last 3.5 million years. Braunschweig: Cramer.

Kutzbach, J.E. & P.J. Guetter 1984. The sensitivity of monsoon climates to orbital parameter changes for 9000 years B.P.: Experiments with the NCAR General Circulation Model. In A.L. Berger (ed.), Milankovitch and Climate, part 2, 801-820. Reidel,

Kutzbach, J.E. & P.J. Guetter 1986. The influence of changing orbital parameters and surface boundary conditions on climate simulations for the past 18,000 years. J. Atmo. Sci. 43: 1726-1759.

Lamb, H.H. 1972. Climate: past, present and future. London: Methuen.

Landlivar, C.B. 1977. El clima y sus caracteristicas en el Ecuador. Xi assemblea general y reuniones panamericanos de cosulta conexas, Quito.

Leyden, B.W. 1985. Late Quaternary aridity and Holocene moisture fluctuations in the Lake Valencia basin, Venezuela. Ecol. 66: 1279-1295.

Livingstone, D.A. 1975. Late Quaternary climatic change in Africa. An. Rev. Ecol. Syst. 6: 249-280.

Liu, K-b, and Colinvaux, P.A. 1985. Forest changes in the Amazon Basin during the last glacial maximum. Nature 318: 556-557.

Meade, R.H., C.F. Nordin, W.F. Curtis, F.M.C. Rodrigues, & J.M. Edmond 1979. Sediment loads in the Amazon River. Nature 278: 161-163.

Milliman, J.D. & H.T. Barretto 1975. Relict magnesium calcite oolite and subsidence of the Amazon shelf. Sedimentology 22: 137-145.

Newell, R.E. 1973. Climate and the Galapagos Islands. Nature 245: 91-92.

Prance, G.T. 1982. Biological Diversification in the Tropics. New York: Columbia.

Projeto Radambrazil 1978. Levantamento de reursos naturais 16: 1-663. Brazilia: DNPM.

Sanford, R.L., J. Saldarriaga, K.E. Clark, C. Uhl & R. Herrera 1985. Amazon rain-forest fires. Science 227: 53-55.

Simpson, B.B. & J. Haffer 1978. Speciation patterns in the Amazonian forest biota. Ann. Rev. Ecol. Syst. 9: 497-518.

Thompson, L.G., E. Mosley-Thompson & B.M. Arnao 1984. El Nino southern oscillation events recorded in the stratigraphy of the tropical Quelccaya ice cap, Peru. Science 226: 50-53.

Wijmstra, T.A., & T. van der Hammen 1966. Palynological data on the history of tropical savannas in northern South America. Leid. Geol. Meded. 38: 71-83.

van der Hammen, T., & E. Gonzalez 1960. Upper Pleistocene and Holocene climate and vegetation of the "Sabana de Bogota" (Colombia, South America). Leid. Geol. Meded. 25: 262-315.

ULRICH RADTKE
Geographical Institute, University of Düsseldorf, FR Germany

14

Marine terraces in Chile (22°-32°S) – Geomorphology, chronostratigraphy and neotectonics : Preliminary results II

ABSTRACT

In the study area of North and Central Chile exist areas of strong and weak tectonic uplift. This is true even in areas which have previously been shown to be stable by means of coastal geomorphology methods and the application of the classic theory of glacial eustatic sea-level variations. The chronostratigraphical coordination with a Quaternary time scale was carried out by using 'absolute' dating methods like Radiocarbon, U-series and Electron-Spin-Resonance (ESR).

1 INTRODUCTION

The ideally built and also well preserved marine terraces of La Serena/Coquimbo have been known to researchers since the famous historical journey of Charles Darwin on the Beagle 1831–1836.

Among the numerous attempts to explain the genesis of marine terraces the eustatic theory gained ground during the 20th century. Fundamental to the considerations was an ocean floor slowly subsiding during the Pleistocene which caused continuous lowering of the sea-level. Superimposed on the slow decrease of the sea-level are the glacial eustatic movements. Since the application of modern physical methods it has become more difficult to apply the classical eustatic hypothesis.

This work on the geomorphology and chronostratigraphy of the Chilean marine terraces (Figure 1) is built on the fundamental and impressing works of Herm (1969) and Paskoff (1970) who derived their results of age and genesis of the marine terraces mainly from altimetrical, geomorphological and paleontological methods and the application of the classic theory of eustatic sea-level oscillations. The renewed research on sequences of terraces described by Herm and Paskoff (e.g.) and the expansion of the study area is the attempt to interprete the known and

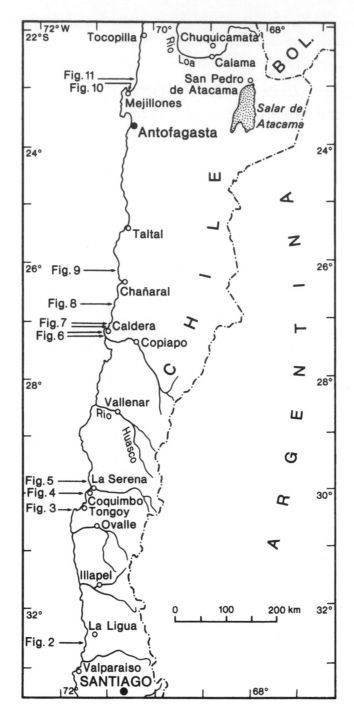

Figure 1. Location of profiles

240

new results in the light of more recent research of sea-level
change and it tries to revise the stratigraphic system of the
marine Quaternary of Chile by Herm and Paskoff (Table 1). In
addition to first results published earlier (Radtke 1985) six
new profiles are presented.

2 STATE OF RESEARCH

The relative oscillations in the sea-level and their influence
on the coastal environment have been used during the last 200
years to corroborate the most divergent geotectonical theories.
However, an understanding of the complex nature of the factors
involved only began in the middle of the 19th century; alas,
in this context the very interesting scientific-historical
discussion of sea-level studies cannot be further elaborated
(Radtke & Ratusny 1987).

It is important that Darwin interpreted the observations made
during his travel along South America concerning the ancient
strandlines of Middle and Northern Chile as indicators of an
uplifting land bound to the volcanisms of the Andes. This
interpretation was a guideline for most Chilean (Brüggen 1950)
and other geologists until our times. It was the geotectonical
point of view expounded by Suess (we owe him the expression
eu-static) which formed the basis of the first works on the
stratigraphy of the Mediterranean marine Quaternary terraces
around 1900; e.g. Deperet correlated the (interglacial) marine
terraces with fluvio-glacial terraces of the northern foreland
of the Alps. Nevertheless his stratigraphic system (Monastirian,
Tyrrhenian, Milazzian, Sicilian, see Table 1) survived. Prominent exponents of the eustatic theory like e.g. Zeuner (1959)
and Fairbridge (1961) applied the classical system without
serious reservations. They concluded that tectonically stable
coasts existed and that therefore the altimetric correlations
of terraces was applicable worldwide. Most research done in the
fifties and sixties was influenced by this theory e.g. the
works of Herm and Paskoff in Chile (Table 1).

For a long time Brüggen's (and Darwin's) ideas of instability
of the Chilean coast since Pliocene had dominated. In contrast
Herm and Paskoff assumed that this uplift was not universal
along the whole coast and that stable sectors might exist.
Instead they favored a eustatic explanation to explain the
marine terraces. Cooke (1964, Huasco) and Segerstrom (1965,
Caldera, Rio Copiapo) believed that most of the terraces were
the result of a single withdrawal, with different pauses in the
general regression. Weischet (1970, Valdivia) accepted the idea
of minor advances and retreats during the Quaternary. Eustatic
sea-level changes have been strongly modified by tectonic movements. During the Quaternary, epeirogenetic uplift has been
dominant north of $40°S$, whereas sinking of the coast has generally prevailed farther south. (Fuenzalida et al. 1965). In

Table 1. The Quaternary marine stages in Central-Northern Chile (according to Herm & Paskoff 1967 and Paskoff 1977)

	Marine Stages		Height a.s.l. (m)	European Stratigraphy	
Holocene	Vega	Middle	2 (14-C: 2.000 B.P.)	Dünkirchen	Postglacial
		Lower	4-5 (14-C: 4.000 B.P.)	Flandrian Calais	
Upper Pleistocene	Cachagua		5-7 (14-C:35.000 B.P.)	Neotyrrhenian	Würm II,III,IV
	Herradura II		15-20	Eutyrrhenian	Würm I
Middle Pleistocene	Herradura I		35-40	Paleotyrrhenian	Riß
	Serena II		75-80	Sicilian	Mindel
Lower Pleistocene	Serena I		120-130	Calabrian	Günz
Upper Pliocene	Coquimbo		200	–	Pre-Günz?

contrast to Paskoff are the findings of Tricart (1965, Rio
Copiapo) who - like Darwin and Brüggen - thought that the
marine terraces had been developped by tectonical uplift. In
addition Stiefel (1974) postulated epeirogenetic oscillations
continuing throughout the Cenozoic. In Pliocene they twice
attained amplitudes of 500 metres and in Quaternary the coast
was raised at the same rate (glacio-eustatic oscillations were
only superimposed). Stiefel correlated his lowest terrace at
100 m a.s.l. with the last interglacial epoch and the second
terrace at 200 m a.s.l. with the penulitmate interglacial.
Even the two following terraces at 300 and 400 m a.s.l. were
dated into the Quaternary. According to him these terrace
sequences of raised fossil beaches were uniform between 27°
and 38°.
  In the following these divergent approaches will be tested
and evaluated in the light of more recent research results of
Quaternary science.

3 RESULTS OF ELECTRON-SPIN-RESONANCE (ESR) AND U-SERIES DATING

The principle of ESR dating is described elsewhere (Hennig &
Grün 1983, Grün 1985). An ESR age is determined according to
the equation:

$$\text{Age (a)} = \frac{\text{accumulated dose (AD)(Gy)}}{\text{annual dose } (D_0)(\text{mGy/a})}$$

The accumulated dose (AD) a sample has received during its geologic age is usually determined by the so-called 'additive-dose'
method: equal, homogenized portions of the sample are irradia-
ted successively with increasing gamma or beta doses and the
ESR intensity is plotted against these doses. The extrapolation
towards zero ESR intensity allows the estimation of the AD.
The annual dose (D_0) is produced by the radioactive elements
in the sample (internal dose-rate) and its surroundings (external dose-rate) (for details see Radtke et al. 1985). Both
dating methods, ESR and U-series, display various interferences.
A main problem in U-series dating mollusc shells in how good
closed system conditions have prevailed over geologic time
(for applicability and validity of the U-series technique see
e.g. Kaufman et al. 1971, Ivanovich & Harmon 1982, Stearns
1984 and Radtke 1986). Recent species often display a low
Uranium-content in the range of 0.5-1 p.p.m., whereas fossil
samples can show a higher U-content (Table 2). U-leaching
cannot be excluded whereas most results predict a post-deposi-
tional U-enrichment. The process leads to apparent younger
U-series ages and a dating attempt will yield the average age
of U-enrichment rather than the age of the shell and correlated
terraces. The derived ages are therefore to be considered as
minimum ages exept when assuming that the accumulation has
occurred very shortly after terrace formation (Radtke et al.

Table 2. U-series ages

Sample	U-238 (p.p.m.)	Th-232 (p.p.m.)	$\frac{U\text{-}234}{U\text{-}238}$	$\frac{Th\text{-}230}{U\text{-}234}$	Th/U-age (ka)
D-691-a	1,22±0,03	0,014±0,006	1,71±0,05	1,29±0,037	∞
D-692-a	1,09±0,017	0,007±0,003	1,22±0,013	0,84±0,02	160^{+10}_{-6}
D-692-b	0,61±0,016	0,215±0,01	1,28±0,05	0,782±0,024	132^{+8}_{-7}
D-694-a	0,79±0,02	0,005±0,004	1,08±0,05	0,98±0,06	$350^{+\infty}_{-110}$
D-698-a	0,85±0,03	0,014±0,006	1,06±0,05	0,726±0,03	134^{+20}_{-8}
D-705-b	0,67±0,02	0,01±0,003	1,07±0,05	0,896±0,06	230^{+60}_{-45}
D-709	1,02±0,031	0,01±0,005	1,47±0,06	0,745±0,035	114^{+11}_{-10}
D-716-b	2,468±0,061	0,034±0,018	1,236±0,025	0,855±0,029	$186,2^{+19,2}_{-16,1}$
D-717-a	0,77±0,02	0,01±0,0005	1,15±0,04	0,792±0,02	154^{+10}_{-6}
D-718-a	1,80±0,05	0,1±0,01	1,17±0,04	0,81±0,04	154^{+24}_{-10}
D-718-b	1,96±0,002	0,055±0,007	1,20±0,02	0,692±0,011	112^{+4}_{-4}

D-729-a	0,89±0,02	0,025±0,05	1,27±0,04	0,67±0,02	106^{+6}_{-4}
D-733	0,275±0,01	0,01±0,01	1,16±0,06	1,006±0,06	$320^{+\infty}_{-70}$
D-784-b	1,087±0,031	0,080±0,016	1,169±0,035	0,899±0,035	$219^{+35,4}_{-26,0}$
D-792-b	0,30±0,06	0,008±0,003	1,28±0,03	0,68±0,04	108^{+7}_{-8}
D-804-a	0,962±0,031	0,017±0,017	1,206±0,043	0,686±0,032	$119,9^{+11,4}_{-10,0}$
D-805-a	5,351±0,057	0,057±0,011	1,162±0,019	0,732±0,022	$136,2^{+8,7}_{-7,9}$
D-806-a	1,372±0,042	0,001±0,016	1,201±0,037	0,688±0,031	$120,5^{+11,1}_{-9,9}$
D-808-a	2,427±0,070	0,067±0,013	1,294±0,032	0,989±0,034	$283,2^{+59,0}_{-38,4}$
D-815-a	1,75±0,05	0,003±0,003	1,00±0,04	0,71±0,03	125^{+15}_{-5}
D-820	2,10±0,03	0,01±0,03	1,55±0,03	0,72±0,02	106^{+6}_{-4}
D-824-b	0,73±0,02	0,04±0,02	1,43±0,03	0,80±0,07	128^{+27}_{-18}
D-926	1,583±0,030	0,050±0,028	1,288±0,018	0,968±0,034	$262,2^{+45,2}_{-32,3}$
D-927	1,033±0,021	0,053±0,011	1,259±0,021	0,922±0,035	$226,6^{+32,2}_{-24,8}$

Table 3. Results of ESR dating; Gy=Gray, D_O=Annual Dose, AD=
Accumulated Dose, ()=External dose-rate estimated, Alpha effi-
ciency=0,1, n.d.=not determined

Sample	D_O (mGy/yr)	AD (Gy)	ESR-age (±20%)(ka)	U-series age (ka)	Species
D-691-a	(1,030)	482,1	≥468,0	∞	Mesodesma
D-691-b	(0,860)	442,5	≥514,5	-	donacium (LAM.)
D-692-a	1,079	150,5	139,4	160	Dosinia sp.
D-692-b	0,974	92,4	94,9	132	
D-694-a	0,804	267,2	≥332,2	350	Protothaca sp.
D-694-b	1,003	204,3	≥203,7	185,8	
D-695-a	2,189	507,0	231,6	-	Mesodesma
D-695-b	2,204	436,0	≥197,8	-	donacium (LAM.)
D-698-a	0,646	51,9	80,3	134	Macrocallista sp.
D-698-b	0,819	73,4	89,6	-	
D-705-b	0,837	61,6	73,6	-	Macrocallista sp.
D-708-a	2,431	303,5	124,8	-	Protothaca thaca
D-708-b	2,576	252,7	99,3	-	(MOLINA)
D-709-b	0,583	47,9	82,1	114	Eurhomalea rufa (LAM.)
D-716-a	1,777	130,5	73,4	-	Protothaca sp.
D-716-b	1,912	154,8	80,9	-	
D-717-a	1,137	220,0	193,0	170	n.d.
D-717-b	1,826	140,3	173,1	-	n.d.
D-718-b	1,752	182,8	104,3	112	Mulinia sp.
D-729-a	1,642	275,5	167,8	106	Mulinia sp.
D-729-b	0,996	111,2	111,9	-	
D-733-a	0,785	268,8	≥342,4	320	n.d.
D-792-a	0,791	67,8	85,7	-	Glycimeris sp.
D-804-a	1,196	99,5	83,2	119,4	Mesodesma sp.
D-8o4-b	1,013	101,4	100,1	-	
D-805-a	2,085	197,4	94,5	135,9	Mesodesma sp.
D-806-a	1,364	152,9	112,1	120,5	Protothaca sp.
D-808-a	1,164	221,3	190,0	282,6	n.d.
D-815	0,974	99,0	101,6	125	Choromytilus chorus (MOLINA)
D-820	1,026	77,8	75,8	106	Eurhomalea rufa (LAM.)
D-824	(143)	124,6	87,1	128	Mulinia sp.

1985). Obviously, a similar problem arises in ESR-dating. The distortion between apparent and real age due to U-accumulation, however, is smaller when changes in the external dose-rate can be excluded, because at the sites of investigation the external radioactivity dominates over the internal radiation sources (Table 3).

4 TERRACE PROFILES

From South to North ten profiles of marine terraces are described.

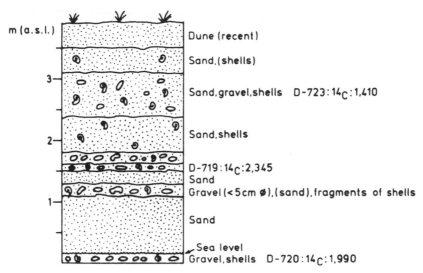

Figure 2. Cachagua (31°35'S)

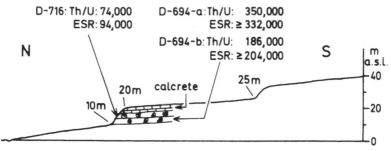

Figure 3. Tangue (P. Aldea/Tongoy)(30°19'S)

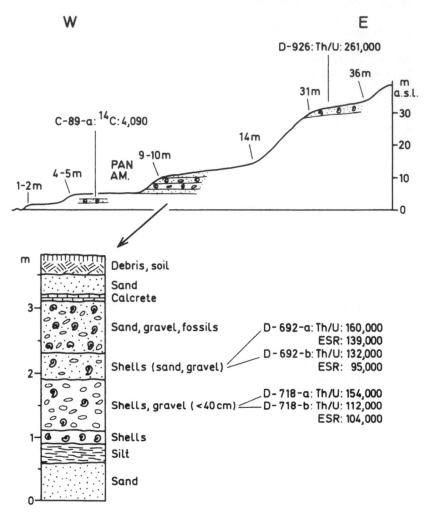

Figure 4. Herradura (29°58'/59'S)

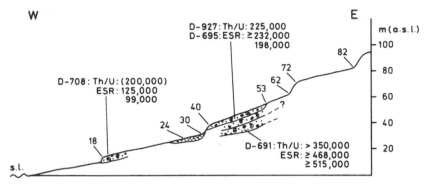

Figure 5. La Serena, Quebr. El Romeral/Quebr. Angostura
(29°48'/49'S)

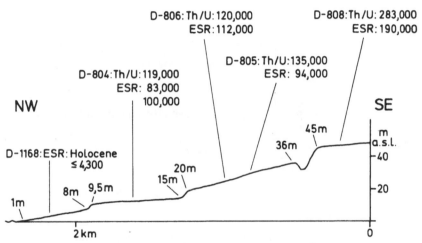

Figure 6. Bahia Inglesa (South of Caldera)(27°07'S)

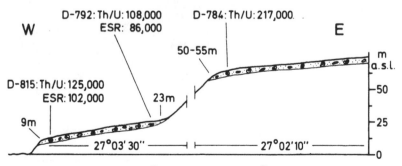

Figure 7. Caldera (27°03'30''S and 27°02'10''S)

249

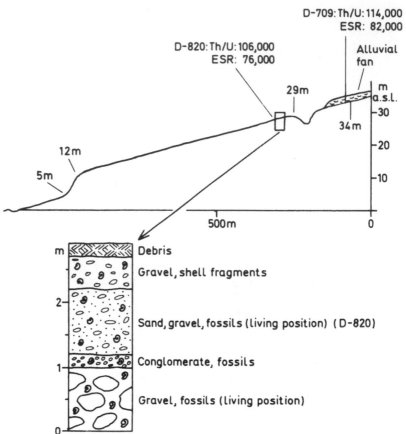

Figure 8. Caleta Obispito (Between Caldera/Chanaral) (26°46'S)

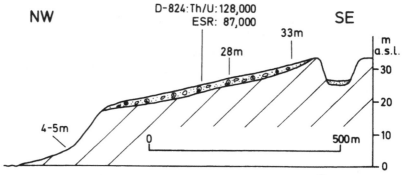

Figure 9. Pan de Azucar (North of Chanaral) (26°08'S)

250

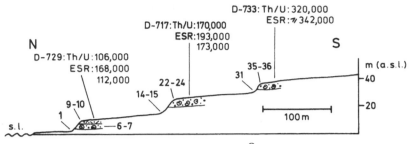

Figure 10. El Rincon, Mejillones (23°04'S)

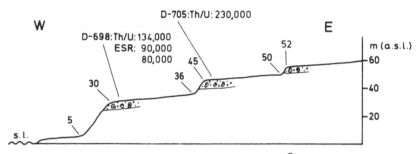

Figure 11. Caleta Playa de Los Hornos (22°55'S)

5 DISCUSSION OF RESULTS

Despite Holocene sediments (Radtke 1985) no signs of a high
sea-level during the last glacial epoch could be found at
Cachagua (Figure 2), the type section of the 'Inner-Würm' high
sea-level in Chile (Table 1, Paskoff 1977). The reconstruction
of this high level as produced by Giresse and Davies as late
as 1980 is definitely based on misinterpretation of absolute
dating results (14-C). Only in coastal regions with an uplift
-rate of 1 m/1000 a as in New Guinea is this 'high-level'
situated above the present sea-level. The proposed Inner-Würm
high sea-level (Paskoff 1973) is also non existent (Radtke
1985) in Coloso (23°45'S, South of Antofagasta). Sediments at
5-7 m a.s.l. could be correlated with the last interglacial
epoch. - Near Puerto Aldea at Tangue/Quebrada Pachingo (Figure
3) a high sea-level of the last interglacial was found at 25 m
a.s.l.; fossils (D-716) belonging originally to the underlying
sediment of the penultimate (?) interglacial are partly reworked.
The fossil coast line of this older sea-level was destroyed
during the maximum of the last interglacial sea-level at 125,000
B.P. - The dating attempts of the type sections Herradura I and
II (Figure 4) still cause difficulties. Most of the absolute

dates permit a coordination with the last interglacial period
(Herradura II, 14 m a.s.l.), although some fossils seem to be
reworked. It is possible that the Herradura I terrace at about
31-36 m a.s.l. does not belong to the penultimate (Table 1),
but may have to be connected with the antepenultimate (?) interglacial. Possibly the terrace of the 200,000 B.P. high-stand
(stage 7) was completely eroded or reworked like the terrace
at P.Aldea. - In contrast to the southern part of the Bay of
La Serena/Coquimbo the northern end is more elevated. The last
interglacial strandline between Quebrada El Romeral and Quebrada
Angostura (Figure 5) was found at 25 m a.s.l., the penultimate
(?) at 40-45 m a.s.l. - Two last interglacial terraces have been
distinguished at Bahia Inglesa (Figure 6), south of Caldera
(9-15 m and 20-36 m a.s.l.). How far this result depends on
dating errors or neotectonical processes needs verification. -
Near Caldera (Figure 7) the sediments of the last interglacial
reached 23 m a.s.l. Whether the fossils at 55 m a.s.l. north
of Caldera (D-784) really belong to the penultimate interglacial
also needs further investigation. - Between Caldera and Chanaral
at Caleta Obispito (Figure 8) sediments of Stage 5 have been
discovered at 34 m a.s.l. Most probably they also exist at
40-43 m a.s.l. - Near Pan de Azucar (Figure 9) the last interglacial was found up to a height of 28 m (fossils) and 33 m
a.s.l. (marine conglomerate). - West of the village of
Mejillones (El Rincon, Figure 10), at the foot of Cerro S. Lucia
which is part of the Horst of Mejillones, three distinct fossil
marine cliffs are visible. The last interglacial sea-level
reached 14 m a.s.l. How far the second (23-31 m) and the third
terrace could be correlated with the penultimate or the antepenultimate interglacial, could not be reliably proved. However,
the sequence of terraces at El Rincon belongs to the very few
localities where U-series and ESR ages yield coherent 'absolute'
ages for the last three interglacial stages 5, 7 and 9 (D-729,
D-717 and D-733). The strong Quaternary uplift of the Horst of
Mejillones, whose fault-line lies only some ten meters West of
El Rincon, most surprisingly has neither tilted nor warped the
three marine terraces. - Some kilometers north of Mejillones at
Caleta Playa de Los Hornos (Figure 11) sediments of the last
interglacial reached 36 m a.s.l.

   Based on the interpretation of the oxygen isotope record
(Shackleton & Opdyke 1973) and the calibration with the K/Ar and
palaeomagnetic method, the knowledge about palaeoclimate and
chronostratigraphy of Quaternary was definitely widened. Investigation of raised coral reef tracts (Barbados, New Guinea; e.g.
Bender et al. 1979, Chappel 1981, Bloom & Yonekura 1985), global
tectonical research and difficulties in correlating mediterranean marine terraces (Hey 1978) have been responsible for the
rejection of the classic eustatic theory. Assuming the relatively constant uplift of Barbados and New Guinea during the
last 700,000 years, it is possible to extrapolate from the altitude of each coral reef tract the related interglacial sea-level

Assuming also a sea-level which was about 5 metres higher than
today at 125,000 B.P., it could be shown that during all interglacial epochs of the Middle and Upper Pleistocene sea-levels -
cum grano salis - oscillated around the present value. Of course,
the assumption of a constant uplift-rate is as questionable as
the postulation of a continuous sedimentation in the deep sea
and the correlated calibration of the oxygen isotope record.
However, it is possible to achive nearly identical results from
Barbados and New Guinea and the deep-sea record with the same
working-hypothesis. Additional research at relatively complete
terrace profiles may help to verify this theory.

As a result of the investigations in Chile the conclusion can
be drawn that the staircases of marine terraces can not be
explained by a more or less continuous sinking of the sea-level
in the Quaternary but only by tectonical movements. The positively dated terrace of the last interglacial is a very useful
indicator for all the neotectonical movements. The different
altitudes of this terrace (5-7 m at Coloso, 14 m at El Rincon
and Herradura, 23 m and 25 m at Caldera and La Serena, 28-33 m
at Pan de Azucar, 36 m at Caleta Hornos and 34-43 m a.s.l. at
Caleta Obispito) give additional validity to the interpretation
of differenciated tectonical movements during the last 125,000
years. Assuming a certain number of faults in discovering and
measuring the exact altitude of the maximum transgression-line,
the variations at the different localities investigated are
nevertheless so extreme that absolute, quantitative statements
have now become possible. Consequently it can be proved that the
postulated stability of certain parts of the coast is fiction.
Even in the area of the type sections at La Serena/Coquimbo
(Serena I, II and Herradura I, II) a different rate of elevation can be found between the northern and southern parts of
the bay. Indeed Paskoff assumed a stronger elevation of the
northern end of the bay but he took the whole only as a local
discrepancy and continued believing in the eustatic theory of
worldwide altimetrically correlated ancient shorelines, although
the supposition of tectonic stability of vast parts of the
coast specifically in the area of the Andean Cordillera with
its high orogenic activity seems unlikely. (Obviously increased
elevations are to be found as on the Mejillones peninsula or
e.g. on the island of Mocha - 25 m/4000 a - in the South of the
Arauco peninsula, but Kaizuka et al. 1973 mistakenly extra-
polated this extreme uplift-rate throughout the whole Pleistocene.)

It is also interesting to note that the geo-tectonical interpretations of Stiefel concerning extended epeirogenetic oscillations seems to be without scientific foundation in Central and
North Chile.

Within the IGCP-200 the investigations about the Chilean
marine terraces strengthen the thesis that sea-level surves of
only regional validity can be provided. Beyond the end of the
classic eustatic theory the study of ancient shorelines is

still of great importance, because former shorelines represent
the position of gravitational equilibrium between lithosphere
and hydrosphere. Shoreline studies have contributed to an better
understanding of deformation of glacio- and hydro-isostatic
origin, thermo- and volcano-isostatic deformation and, especially important for Chile, crustal block movements at subduc-
tion zones. Investigations of ancient shorelines seem to be
very effective means of improving the understanding of the rheo-
logical properties of the Earth and a further aim is the reconstruction of accurate sequences of palaeogeoidal profiles
(Pirazzoli & Grant 1986).

ACKNOWLEDGEMENTS

I should like to thank Dr. A. Mangini, Heidelberg and Dr. R.
Hausmann, Köln for the U-series age determinations. I also want
to thank Mr. O. Katzenberger for the Fission Track analysis of
some shells and Dr. H. Pietzner for the U-, Th- and K-analysis
of the sediments. I am grateful to Prof. Dr. D. Herm, München,
for paleontological determination of some fossils. This research
project was financially supported by the DFG (Deutsche For-
schungsgemeinschaft).

REFERENCES

Bender, M.L., Fairbanks, R.G., Taylor, F.W., Matthews, R.K.,
    Goddard, J.K. & W.S. Broecker 1979. Uranium-series dating of
    the Pleistocene reef tracts of Barbados, West Indies.
    Geol. Soc. Am. Bull. I, 90(I):577-594.
Bloom, A.L. & N. Yonekura 1985. Coastal terraces generated by
    sea-level change and tectonic uplift. In J.Woldenberg (ed.),
    Models in Geomorphology, p.139-154. Boston: Allen & Unwin.
Brüggen, M.J. 1950. Fundamentos de la Geologia de Chile.
    374 pp. Santiago: Instituto Geografico Militar.
Butzer, K.W. 1983. Global Sea Level Stratigraphy: An Appraisal.
    Quat. Sci. Rev. 2:1-15.
Chappel, J. 1981. Relative and average sea level changes, and
    endo-, epi- and exogenic processes on the earth. In
    I.Allison (ad.), Sea level, ice and climatic change, IAHS
    Publ. 131:411-430, Washington.
Clark, J.A. 1980. A numerical model of worldwide sea level
    changes on a viscoelastic earth. In N.A.Mörner (ed.), Earth
    Rheology, Isostasy and Eustasy, p.525-534. Chichester:
    Wiley-Interscience.
Cooke, R.U. 1964. Les niveaux marins des baies de La Serena et
    de l'Huasco. Bull. Assoc. Géogr. Franc. 321:19-37.
Cronin, T.M. 1983. Rapid Sea Level and Climatic Changes:
    Evidence from Continental and Island Margins. Quat. Sci. Rev.
    1:177-214.

Fairbridge, R.W. 1961. Eustatic changes in sea level. Physics and Chemistry of the Earth 4:99-185.

Fuenzalida, H., Cooke, R.U., Paskoff, R., Segerstrom, K. & W.Weischet 1965. High stands of Quaternary sea-level along the Chilean coast. Geol. Soc. Am., Spec. Paper 84:473-496.

Giresse, P. & O.Davies 1980. High sea levels during the last Glaciation. One of the most puzzling problems of sea-level studies. Quaternaria 22:211-236.

Grün, R. 1985. Beiträge zur ESR-Datierung. Sonderveröff. Geol. Inst. Univ. Köln 59:1-157

Hennig, G.J. & R.Grün 1983. ESR dating in Quaternary Geology. Quat. Sci. Rev. 2:157-238.

Herm, D., Paskoff, R. & J.Stiefel 1966. Premiéres observations sur les alentours de la baie de Tongoy (Chili). Bull. Soc. Géol. France 7, VIII:21-24.

Herm, D. & R.Paskoff 1967. Vorschlag zur Gliederung des marinen Quartärs in Nord- und Mittel-Chile. N. Jahrb. Geol. Paläont., Monatshefte 10:577-588.

Herm, D. 1969. Marines Pliozän und Pleistozän in Nord- und Mittel-Chile unter besonderer Berücksichtigung der Entwicklung der Mollusken-Faunen. Zitteliana 2:1-159.

Hey, R.W. 1978. Horizontal Quaternary shorelines of the Mediterranean. Quat. Res. 10:197-203.

Ivanovich, M. & S.Harmon 1982. Uranium series disequilibrium: Applications to environmental problems. Oxford.

Kaizuka, S., Matsuda, S., Nogami, M. & Yonekura, N. 1973. Quaternary tectonic and recent seismic crustal movements in the Arauco Peninsula and its Environs, Central Chile. Geogr. Rep. Tokyo Metropolitan University 8:1-49.

Kaufman, A. 1986. The distribution of 230-Th/234-U ages in corals and the number of Last Interglacial high-sea stands. Quat. Res. 25(1):55-62.

Kaufman, A., Broeker, W.S., Ku, T.L. & Thurber, D.L. 1971. The status of U-series methods of mollusk dating. Geoch. Cosmoch. Acta 35:1155-1183.

Mörner, N.A. 1981. Eustasy, Paleoglaciation and Paleoclimatology. Geol. Rundschau 70:691-702.

Paskoff, R. 1970. Le Chili semi-aride. Recherches géomorphologiques. 420pp., Bordeaux: Biscaye Fréres.

Paskoff, R. 1973. Radiocarbon dating of marine shells taken from the north and central coast of Chile. IX. International INQUA Congress Abstracts, Christchurch, p.281-282.

Paskoff, R. 1977. Quaternary of Chile: The State of Research. Quat. Res. 8:2-31.

Pirazzoli, P.A. & Grant, D.R. 1986. Lithospheric deformation deduced from ancient shorelines. Litoralia, Newsletter 12:8-11.

Radtke, U. 1985. Chronostratigraphie und Neotektonik mariner Terrassen in Nord- und Mittelchile - Erste Ergebnisse. IV Congreso Geológico Chileno - Universidad del Norte - Antofagasta, Tomo III:436-457.

Radtke, U. 1986. Value and risk of radiometric dating of shorelines-geomorphological and geochronological investigations in Central Italy, Eolian Islands and Ustica (Sicily). In: A. Ozer & C. Vita-Finzi (eds.), Dating Mediterranean Shorelines. Z. Geomorph. N.F. Suppl. Bd. 62:167-181.

Radtke, U. 1987. Palaeo Sea Levels and Discrimination of the Last and the Penultimate Interglacial Fossiliferous Deposits by Absolute Dating Methods and Geomorphological Investigations - Illustrated from Marine Terraces in Chile. Berliner geographische Studien, 25:313-342.

Radtke, U., Hennig, G.J., Linke, W. & Müngersdorf, J. 1981. 23o-Th/234-U- and ESR-dating problems of fossil shells in Pleistocene marine terraces (Northern Latium, Central Italy). Quaternaria 23:37-50.

Radtke, U., Mangini, A. & Grün, R. 1985. ESR dating of marine fossil shells. Nuclear Tracks 10(4-6):879-884.

Radtke, U. & Ratusny, A. 1987. Meeresspiegelschwankungen im Quartär - forschungsgeschichtlicher Rückblick und neue Perspektiven. Berliner Geographische Studien (in press).

Segerstrom, K. 1965. Dissected gravels of the Rio Copiapo Valley and adjacent coastal area, Chile. U.S. Geol. Survey Prof. Paper 525-B:117-121.

Shackleton, N.J. & Opdyke, N.D. 1973. Oxygen Isotope and Palaeomagnetic Stratigraphy of Equatorial Pacific Core V-28-238: Oxygen Temperatures and Ice Volumes on a 10^5 and 10^6 Year Scale. Quat. Res. 3:39-55.

Stearns, C.E. 1984. Uranium-series dating and the history of sea level. In: W.C. Mahaney (ed.), Quaternary Dating Methods, p.53-66, Amsterdam:Elsevier.

Stiefel, J. 1974. Zur tektonischen Interpretation jungkänozoischer Sedimente und Landformen in der Küstenzone Mittelchiles. Geotekt. Forsch. 46:70-194.

Tricart, J. 1965. Algunas observaciones geomorfológicas sobre las terrazas del Río Copiapo. Inf. Geogr. 15:45-60.

Weischet, W. 1970. Chile, seine länderkundliche Individualität und Struktur. 618pp., Darmstadt: Wissenschaftliche Buchgesellschaft.

Zeuner, F.E. 1959. The Pleistocene Period. 2nd edn., London: Hutchinson.

KENITIRO SUGUÍO & MOYSÉS GONZALES TESSLER
Universidade de São Paulo, Brazil

15

Characteristics of a Pleistocene nearshore deposit: An example from southern São Paulo State coastal plain

ABSTRACT

Cananéia Formation, occurring almost continuously at least in the southern half of the State of São Paulo coastal plain, is overlying continental Tertiary deposits of Pariquera Açu Formation or crystalline rocks of Precambrian age. Its basal portion, in general, with a sandy-clayey composition is superimposed by essentially sandy sediments, both representing the Pleistocene Cananéia Transgression (=Sangamon) in the area.

A detailed study of this formation in Comprida and Cananéia Islands, as well as in neighboring coastal plain in the continent, allowed us to recognize an assemblage of sedimentary structures representative of shallow marine depositional environments.

The fine sands with heavy mineral laminations and sandy-clayey intercalations, exhibiting wavy, lenticular and flaser beddings associated with intensive bioturbations and load-casts, are probably related to upper shoreface. These sediments are overlain by purer fine sands with parallel horizontal to sub-horizontal laminations and low-angle cross-beddings and mudcracks, representing a shallower sub-environment subjected to periodical subaerial exposures in a foreshore, indicate the end of the Cananéia Transgression.

RESUMO

A Formação Cananéia, que ocorre mais ou menos continuamente pelo menos na metade sul da planície costeira do Estado de São Paulo, é superposta aos depósitos terciários continentais da Formação Pariquera Açu ou as rochas cristalinas de idade pré-cambriana. A sua porção basal apresenta, em geral, uma constituição areno-argilosa, sendo encimada por depósitos essencialmente arenosos, ambos representando a Transgressão Cananéia (= Sangamoniana) de idade pleistocênica na área.

Um estudo detalhado desta formação nas ilhas Comprida e Cananéia, bem como na planície costeira adjacente do continente, permitiu-nos reconhecer um conjunto de estruturas sedimentares representativas de ambientes deposicionais de mar raso.

As areias finas com laminações de minerais pesados e intercalações areno-argilosas, exibindo acamamentos ondulados, lenticulares e flaser associados a intensas bioturbações e estruturas de sobrecarga, são provavelmente relacionadas a face praial superior. Estes sedimentos são superpostos por areias finas mais puras com laminações horizontais e sub-horizontais paralelas e com estratificações cruzadas de baixo ângulo e gretas de contração, representando um sub-ambiente mais raso submetido a exposições sub-aéreas periódicas em uma antepraia, indicativas de fim da Transgressão Cananéia.

INTRODUCTION

Cananéia-Iguape coastal plain, situated in the southern part of the State of São Paulo, has an area of about 2,500 km^2, with a roughly half-moon shape of 130 x 40 km (Figure 1). Its northeastern and southwestern extremities are constituted by Precambrian crystalline rock headlands which reach the sea. This area is dissected by lagoonal channel systems delineating three islands, the first two essentially formed by Quaternary deposits (Cananéia and Comprida Islands) and the last one by Precambrian crystalline rocks (Cardoso Island).

Comprida Island is a barrier-island mostly related to sea-level fluctuations of the last 7,000 years (Martin and Suguic, 1978). It is about 70 km long and 3 - 5 km wide. A lagoonal channel 400 to 1,200 m wide, locally named "Mar Pequeno", separates this island from the continent. Southward, "Mar Pequeno" is divided into two lagoonal channels ("Mar de Cubatão" and "Mar de Cananéia"), around the Cananéia Island.

In Cananéia Island and in the neighboring coastal plain in the continent, the Quaternary shallow-marine and lagoonal (or bay bottom) sediments of Pleistocene Cananéia Formation are dominant. This formation was, for the first time,

258

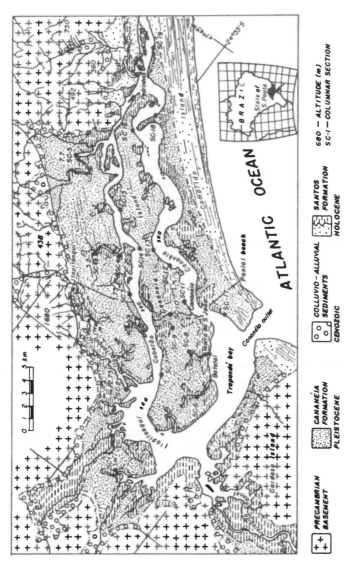

Figure 1. Geologic map of the studied area.

recognized and described by Suguio and Petri (1973). Its minimum age was obtained by Suguio and Martin (1978a), by dating wood fragments contained in basal clayey-sandy sediments. This formation has been correlated with the Pleistocene deposits in the State of Bahia coastal plains, dated by Martin **et al.** (1982) as 120,000 years BP.

The surface of Cananéia Formation was intensively dissected, during the following low sea-level period (Northern Hemisphere glacial period), by drainage net established at that time. During the last 6 - 7,000 years, sea-level surpassed the present sea-level, when the eroded portions of the Cananéia Formation have been occupied by the Holocene Santos Formation.

Suguio and Martin (1978a, b) reported the results of a detailed study, based on the interpretation of aerial photos, field surveys, and radiocarbon dating of these Quaternary sediments. They established the following evolutive model (Figure 2), valid for the Cananéia-Iguape coastal plain:

First stage - During the maximum of Cananéia Transgression the sea reached the foot of the Serra do Mar. In this period, clayey-sandy and transgressive marine sands covered the Pariquera Açu Formation.

Second stage - With the regressive phase, beach-ridges began to be deposited on the top of sandy sediments.

Third stage - During this phase sea-level was always lower than the present one (about 18,000 years BP the sea-level was about - 110 m below present level), and the rivers deeply eroded the sedimentary deposits of Cananéia Transgression. Valleys were formed similar to those in the Barreiras Formation, observed along the coastline of the State of Bahia.

Fourth stage - During the last transgressive phase the sea encroached upon lower zones at the beginning forming an extensive lagoonal system where clayey-sandy deposits, frequently very rich in organic matter, were deposited. In the meantime, the sea eroded the higher-lying parts of Cananéia Formation and redeposited the eroded sands to form the Holocene sandy marine deposits.

Fifth stage - During the return of the sea-level toward the present level, regressive beach-ridges were formed. The fluctuations of the sea-level during the final part of Santos Transgression produced several generations of beach-ridges. Thus, on Comprida Island we can see at least two generations of beach-ridges separated by a rather swampy, low-lying zone that can be followed over aproximately 50 km.

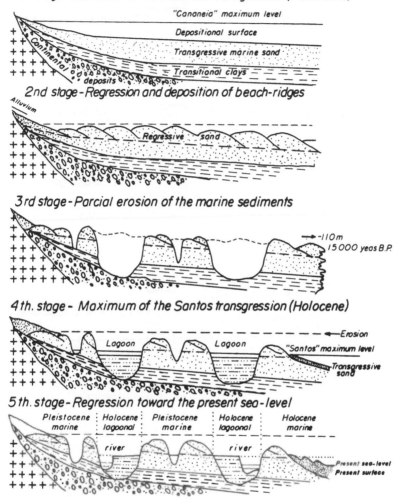

1 st. stage-Maximum of the Cananéia transgression (Pleistocene)

"Cananeia" maximum level

Depositional surface

Transgressive marine sand

Transitional clays

Continental deposits

2nd stage-Regression and deposition of beach-ridges

Alluvium

Regressive sand

3rd stage-Parcial erosion of the marine sediments

→ -110m
15 000 yeas B.P.

4th. stage - Maximum of the Santos transgression (Holocene)

← Erosion

Lagoon   Lagoon

"Santos" maximum level

Transgressive sand

5th. stage-Regression toward the present sea-level

| Pleistocene marine | Holocene lagoonal | Pleistocene marine | Holocene lagoonal | Holocene marine |

river   river

Present sea-level
Present surface

Figure 2. Evolutive stages proposed to explain the origin of
the Cananéia-Iguape coastal plain (Suguio & Martin, 1978a).

CANANÉIA FORMATION SEDIMENTARY STRUCTURES

Outcrops of Cananéia Formation, whose summits are situated
about 6 to 8 m above present high-tide level, are very
frequently found in the studied area. They are always
associated with erosional margins of present lagoonal channels.
Many kinds of sedimentary structures have been observed by

261

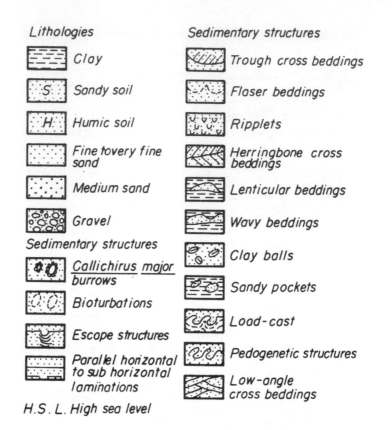

Lithologies

≡≡≡ Clay

S Sandy soil

H Humic soil

Fine tovery fine sand

Medium sand

Gravel

Sedimentary structures

Callichirus major burrows

Bioturbations

Escape structures

Parallel horizontal to sub horizontal laminations

H.S.L. High sea level

Sedimentary structures

Trough cross beddings

Flaser beddings

Ripplets

Herringbone cross beddings

Lenticular beddings

Wavy beddings

Clay balls

Sandy pockets

Load-cast

Pedogenetic structures

Low-angle cross beddings

Figure 3. Legend for lithologies and sedimentary structures.

previous authors but they have never been studied in detail.
Thirty six local columnar sections were measured and described
in detail which allowed us to recognize the following features
(Figure 3):

1. Biogenic structures - **Callichirus major** burrows, bioturbations (distinct and indistinct mottled structures), and
escape structures. They have been described and interpreted
previously by Suguio and Martin (1976), Suguio **et al.** (1984),
and Rodrigues **et al** (1985).

2. Hydrodynamic structures - parallel horizontal and subhorizontal laminations, trough cross-beddings, flaser beddings,
ripple-drift cross-laminations, herringbone cross-bedding,
lenticular beddings, wavy beddings and low-angle cross-
beddings. These and some other miscellaneous structures, like
load casts, clay galls and clay balls, mudcracks have been

262

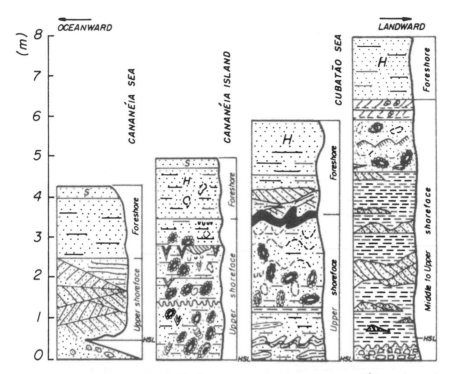

Figure 4. Integrated columnar sections of the Cananéia coastal plain showing the most important lithologies and sedimentary structures.

observed in Cananéia Formation. Some of them have been reported previously by Suguio and Petri (1973), and the meaning of hydrodynamic structures are explained in many text books on sedimentary structures.

Many of these sedimentary structures (**Callichirus major** burrows, flaser beddings, etc.) are very useful for the sedimentary environmental interpretation of Cananéia Formation.

COLUMNAR SECTIONS

Thirty six local columnar sections, representing outcrops of Comprida and Cananéia Islands, as well as the neighboring coastal plain in the continent, have been integrated into four composite sections, which give an idea of the general trend of facies changes from the open-sea toward the continent (Figure 4).

COLUMNAR SECTION REPRESENTING COMPRIDA ISLAND

The outcrop representative of the Comprida Island inner portion is located in southernmost extremity of the island (Picarro do Morrete) and is only 4 m high, because it is a wave-cut terrace built on Cananéia Formation.

It is entirely composed of fine to very fine sand without sandy-clayey layers. Abundant **Callichirus major** burrows, observed at present high-tide level, have been studied in detail by Suguio **et al.** (**op. cit.**) and Rodrigues **et al.** (**op. cit.**). Toward the top there is a sandy layer about 2 m thick with abundant herringbone cross-beddings, followed by sands with parallel horizontal to sub-horizontal laminations and low-angle cross-laminations.

COLUMNAR SECTIONS REPRESENTING CANANÉIA ISLAND

Cananéia Island outer portion - The most representative outcrop of this part of the island is found northward of Hotel Glória (Cananéia town). About lower 2/3 of this 5 m high outcrop constitutes the bed rich in **Callichirus major** burrows and escape structures, which are completely obliterating the primary hydrodynamic sedimentary structures. These tubes may be observed in longitudinal, transversal and diagonal sections, which give a good idea of their morphology. Horizontal parallel laminations, ripple-drift cross-laminations and trough cross-beddings have observed in some places. About 1.5 m of its uppermost portion is characterized by parallel horizontal laminations and indistinct mottled structures.

Clayey-sandy intercalations are not observed in this portion of Cananéia Island, but they are present in some places in the inner part of the island, like in a sand quarry near the ferry-boat pier to the continent.

Cananéia Island inner portion - It is well represented by outcrops situated in front of the Iririaia-Açu River mouth. The outcrops are 5 to 6 m high, and about 1.5 m basal portion is very rich in load-casted clayey-sandy intercalations. This is followed toward the top by 2 m thick sand bed characterized by frequent **Callichirus major** burrows, indistinctly mottled structures and common flaser beddings. The top of these outcrops are also characterized by parallel horizontal to sub-horizontal and some trough cross-beddings.

264

COLUMNAR SECTIONS REPRESENTING THE CONTINENT

As the outcrops heights and the level of **Callichirus major** burrows increase landward, here are situated the highest outcrops in the studied area, wll represented by the one situated in the left margin of the Iriaia-Açu River.

About 6 m of its basal portion is dominantly constituted by greenish-grey argillaceous sediments intercalated by fine to very fine sand layers. These sands occur in association with wavy and lenticular beddings and some sandy patches filling organism burrows forming distinct mottled structures. In some places of the coastal plain in the continent, as in the Itapitangui River mouth, Cananéia Formation is underlain by oxidized pebbly muds, probably representing Pariquera Açu Formation, continental deposits, which according to Sundaram and Suguio (1985), is of Pliocene age.

About 1.5 m thick sandy layer with **Callichirus major** burrows are, in the area, found about 5 m above the present high-tide level.

The uppermost 1.5 m is formed of fine to very fine sand with parallel horizontal to sub-horizontal laminations and low-angle cross-laminations.

DEPOSITIONAL ENVIRONMENTS

According to Petri and Suguio (1973), Cananéia Formation represents a transgressive episode, beginning with transitional (brackish water) clayey-sandy sediments, followed by shallow-marine sands in its upper portion. This interpretation was based on microorganisms (foraminifera and diatoms) and grain size analysis.

Lithological characteristics, mostly sedimentary structures here analysed in some detail, are very important to understand the depositional environments of the Cananéia Formation. The lower portion characterized by an abundant hydrodynamic and biogenic structures could be suitably attributed to upper shoreface sub-environment. Lenticular, wavy and flaser beddings are suggestive of an environment where tidal currents played a very important role during sedimentation. The heights of the levels with abundant **Callichirus major** burrows, increasing from present shoreline toward the continent, and the anomalously higher density of burrows found by Suguio **et al.** (**op. cit.**) in "Costeira da Barra" area (Ribeira de Iguape

River) suggest sea-level rising during its deposition. About 1.5 to 2 m uppermost portion of Cananéia Formation outcrops, with only few structures and in some places exhibiting evidence of reworking by wind, could have been deposited in a foreshore sub-environment. The sedimentary structure most typical of foreshore is the low-angle cross-bedding (Thompson, 1937; McKee, 1957).

Therefore, sediments deposited in deepest water within the sequence must be represented between the above mentioned sub-environments, but this evidence is not clear. On the other hand, the striking difference between the columnar sections representing Cananéia Island, and the neighboring coastal plain in the continent, could indicate a much quieter water (lagoonal channel?) during the deposition of sediments outcropping in the Iririaia Açu River. This fact would suggest that Cananéia Island was a barrier-island during the Pleistocene originating a protected area at its backside similar to the present situation.

ACKNOWLEDGEMENTS

This work has been made possible thanks to financial support from FAPESP (Fundação de Amparo a Pesquisa do Estado de São Paulo) for field surveys (Processo 84/0271-0).

REFERENCES

Martin, L.; Bittencourt, A.C.S.P. & Vilas-Boas, G.S. 1982. Primeira ocorrência de corais pleistocênicos da costa brasileira: Datação do máximo da penúltima transgressão. **Ciencia da Terra**, 3:16-17, Salvador (BA).
Martin, L. & Suguio, K. 1978. Ilha Comprida: Um exemplo de ilha-barreira ligado a flutuações do nível marinho durante o Quaternário. **XXX Congr. Bras. Geol. Anais** 2:905-912, Recife (PE).
McKee, E.D. 1957. Primary structures in some recent sediments. **Amer. Assoc. Petrol. Geol. Bull.** 41:1704-1747.
Rodrigues, S. de A.; Suguio, K. & Shimizu, G.Y. 1984. Ecologia e paleoecologia de **Callichirus major** Say (1818) (Crustacea, Decapoda, Thalassinidea). **An. Sem. Reg. Ecol.**, 4:499-519, São Carlos (SP).

Suguio, K. & Martin, L. 1976. Presenca de tubos fósseis de
Callianassa nas formações quaternárias do litoral paulista
e sua utilização na reconstrução paleoambiental. Bol. IG,
Inst. Geocienc., USP, 7:17-26. São Paulo (SP).

Suguio, K. & Martin, L. 1978a. Quaternary marine formations
of the State of São Paulo and southern Rio de Janeiro. 1978
International Symposium on Coastal Evolution in the
Quaternary Special Publication N° 1: 55p., São Paulo (SP).

Suguio, K. & Martin, L. 1978b. Mapas geológicos das planícies
costeiras quaternárias do Estado de São Paulo e sul do Rio
de Janeiro (1:100.000). DAEE/SOMA, São Paulo.

Suguio, K. & Petri, S. 1973. Stratigraphy of the Iguape-Cana
néia lagoonal region sedimentary deposits, São Paulo, Brazil.
Part I: Field observations and grain size analysis. Bol. IG,
Inst. Geocienc., USP, 4:1-20, São Paulo (SP).

Suguio, K.; Rodrigues, S. de A.; Tessler, M.G. & Lambooy, E.E.
1984. Tubos de ophiomorphas e outras feicoes de bioturbação
na Formação Cananéia, Pleistoceno da planície costeira
Cananéia-Iguape, SP. In: Lacerda, L.D. et al. (organizado-
res). Restingas: origem, estrutura, processos, 111-122,
Niterói (RJ).

Sundaram, D. & Suguio, K. 1985. Nota preliminar sobre uma
assembléia mioflorística da Formação Pariquera Açu, Estado
de São Paulo. VIII Congr. Bras. Paleont. (1983), MME-DNPM,
Série Geologia N° 27, Paleont./Estrat. N° 2:503-505, Rio de
Janeiro (RJ).

Thompson, W.O. 1937. Primary structures of beaches, bars and
dunes. Geol. Soc. Amer. Bull., 48:723752.

JORGE O.RABASSA
CADIC-CONICET, Tierra del Fuego, Argentina

16

The Holocene of Argentina: A review

ABSTRACT

The Holocene of Argentina between lat. 25°-56°S is reviewed
according to the present knowledge of four large regions:
1) the Buenos Aires Pampas; 2) Mesopotamia and Chaco; 3) the
Central Andes; and 4) the Patagonian and Fuegian Andes.
   The Holocene of the Pampas is characterized by the
accumulation of friable, loose loess on the divides and fine
grained alluvial sediments along the valleys, two major
pedogenetic episodes and the extinction of the fossil megamammal fauna. The climate has been estimated as "wet-warm".
Palynological evidence suggests that this was reversed towards
the end of the Flandrian transgression, leading to drier,
steppe-like conditions, until present times. The Mesopotamian
region shows a complex evolution during the Quaternary, through
the interaction of four major environments: the Paraná River,
the western alluvial fans, the Brazilian Shield and the Pampean
Region. Climate has been wet and warm, but between 3000-1000 yr.
BP more arid conditions developed. Neotectonics have been very
active. Glacier fluctuations in the Central and Patagonian
Andes, together with volcanic and pyroclastic eruptions, are
the most significant Holocene episodes in these regions. In the
Central Andes, only two major Neoglacial expansions have been
clearly recorded so far. Further south, the Patagonian Andes
show, instead, evidence of up to twelve Holocene readvances,
peaking at 8600, 8200, 5200, 4600, 3300, 2700, 2200, 1300, 900,
600, 300 and 150 yr. BP.

RESUMEN

El Holoceno de Argentina, entre las latitudes 25°-56°S, es
considerado de acuerdo al conocimiento actual en cuatro grandes
regiones: 1) las Pampas de Buenos Aires; 2) Mesopotamia y
Chaco; 3) los Andes Centrales; y 4) los Andes de Patagonia
y Tierra del Fuego.

El Holoceno de las Pampas está caracterizado por la acumulación de loess friable y suelto sobre las divisorias y sedimentos aluviales de grano fino, a lo largo de los valles, dos
episodios pedogenéticos mayores y la extinción de la fauna
fósil de megamamíferos. El clima ha sido estimado como "húme-
do-cálido". La evidencia palinológica disponible sugiere que
estas características fueron revestidas hacia el final de la
transgresión Flandriana, generando condiciones climáticas más
secas, esteparias, hasta la actualidad.

La región mesopotámica muestra una evolución compleja durante
el Cuaternario, debido a la interacción regional de cuatro
ambientes principales: el Río Paraná, los abanicos aluviales
occidentales, el escudo brasiliano y la región pampeana. El
clima era húmedo y cálido, pero entre 3000 y 1000 a AP, se
desarrollaron condiciones más áridas. La neotectónica ha sido
muy activa. Las fluctuaciones glaciales en la Cordillera de
los Andes, junto a las erupciones volcánicas y piroclásticas,
son los episodios holocenos más significativos. En los Andes
Centrales, sólo dos expansiones neoglaciales han sido registradas claramente hasta el presente. Más al sur, los Andes
Patagónicos muestran, en cambio, evidencia de hasta doce re-
avances holocenos, con culminaciones hacia 8600, 8200, 5200,
4600, 3300, 2700, 2200, 1300, 900, 600, 300 y 150 años AP.

INTRODUCTION

Since the recent, though already classical, paper by Fidalgo
and Tonni (1982), significant work has been done concerning
the Holocene of Argentina. This paper is an attempt to present
an up-dated review of such knowledge.

Argentina occupies, with Chile, the southernmost part of
South America, Argentina faces towards the Atlantic Ocean
and Chile does it so towards the Pacific. Both countries
share a 5000 km long, common back bone: the Andean Cordillera
(Figure 1).

Figure 1. Location map.

Such enormous stretch of land shows a highly-varying geographical and climatic setting from north to south. Very warm, either extremely dry (Atacama Desert) to highly wet (NE Argentina), tropical lands in the northern part give way to Mediterranean type regions (Central Chile), arid and semi-arid piedmonts (Western Argentina) and mild loess prairies (the Argentina Pampas), around lat. 30°-38°S. Southwards, a strong contrast develops between the very wet mountains of the Patagonian Andes, which bear the southern beach rain forest, to the cold and dry steppes of the Patagonian tablelands that extend all the way to the Atlantic coast.

In such variety of environmental conditions, it seems clear that the characteristics, genesis and extent of the Holocene deposits of Argentina present a complex and diverse regional distribution.

Four areas will be considered under this regional approach: 1) the Pampean Region, characterized by the thick accumulation of loess; 2) the Northeastern Region, related with tropical and subtropical environments; 3) the Central Andes, high mountain chains and extended piedmont areas, under arid climatomorphogenesis; and 4) the Patagonian Andes, where at least three major Neoglacial expansions are well represented. All four regions share a main, common characteristic: they all depict the environmental impact of climatic change, both during the Pleistocene-Holocene transition and throughout the Holocene epoch.

1 THE HOLOCENE OF THE BUENOS AIRES PAMPEAN REGION

The Pampean Region of Buenos Aires Province extends between lat. 33°S and lat. 38°S, approximately. It comprises more than 300,000 sq. km of lowlands and grassy, flat prairies with extremely low gradients and two low-altitude hilly chains, Sierras de Tandil y Sierra de la Ventana. Most of this region is covered by thick Late Pleistocene and Holocene loess deposits (Figure 2).

The climate is mild and wet, with cool winters (though it hardly freezes and never snows) and very warm summers. The dominant winds have long been from the W and SW, and they have been responsible for the transport and accumulation of the silt and dust of arid and volcanic origin from the Western and Patagonian deserts into the Pampas.

The region has been tectonically stable for long time. Two positive areas, the sierras, and one major tectonic trough, Río Salado Depression, dominate the flat landscape. No neo-tectonic activity has been definitively proved yet, though it has been postulated by Pasotti (1974) and Iriondo (1984) for other areas immediately north of here.

The study of the Pampean Holocene effectively started with the outstanding work of Ameghino (1889) and for long, it remained based on the study and interpretation of fossil mammal remains.

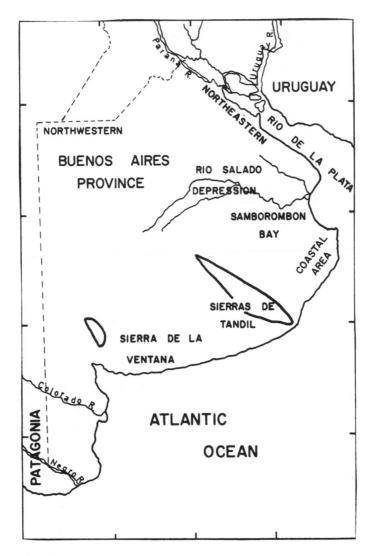

Figure 2. Map of Buenos Aires Province.

More recently, the definition of litho - and pedo-
stratigraphic units by Fidalgo and collaborators in the Salado Depression (see, for example, Fidalgo 1979; Fidalgo and
Tonni 1982) provided a basic framework for the understanding
of spatial relationships, chronology and climatic change.

Further studies by several authors have generally supported
Fidalgo's stratigraphic schemes and correlated it with units
in other areas (Table 1).

Table 1. Stratigraphic correlation of the Holocene of the Buenos Aires Pampean Region.

AGE	RIO SALADO DEPRESSION [1]		COASTAL AREA [2]		NORTHEAST [3]		NORTHWEST [4]	TANDIL [5]		VENTANA [6]	
	FLUVIAL	AEOLIAN	MARINE	COASTAL SANDS	FLUVIAL	AEOLIAN	AEOLIAN	FLUVIAL	AEOLIAN	FLUVIAL	AEOLIAN
0	ALLUVIUM		LA POSTRERA		ALLUVIUM		SERE Mb	ALLUVIUM	III	CHACRA / LA BLANQUEADA Fm	MATADERO / SALDUNGARAY Fm
	SOIL		SOIL		SOIL		ASH AD 1932	SOIL	SOIL	SOIL	SOIL
1000		LA POSTRERA Fm	LAS ESCOBAS Fm	CANAL 18 Mb	RIO SALADO Mb	LA POSTRERA Fm	SANTA INES Mb	UPPER SILTY Mb	II	UPPER SILTY-SANDY Mb	SAAVEDRA Fm
3000	RIO SALADO Mb			CERRO DE LA GLORIA Mb	SOIL		LAS LILAS Fm	MIDDLE SANDY Mb	SOIL I	MIDDLE SANDY Mb	UPPER MEMBER
	SOIL		SOIL						LAS ANIMAS Fm	AGUA BLANCA Fm	SOIL
	GUERRERO Mb		DESTACAMENTO RIO SALADO Fm		GUERRERO FORMATION		LA CABAÑA SOIL	TANDILEOFU Fm			MIDDLE Mb
7800	LUJAN FORMATION	PAMPIANO FORMATION	PAMPIANO FORMATION		LUJAN FORMATION / PAMPIANO FORMATION		CARLOS TEJEDOR Fm				
10000											

HOLOCENE

LATE PLEISTOCENE

References:
1. Fidalgo and Tonni (1982)
2. Fidalgo (1979)
3. Fidalgo and Tonni (1982)
4. Dillon, A. (unpublished)
5. Rabassa (1974)
6. Rabassa (1985)

274

The Holocene lithogenesis in the lowlands is characterized
by the accumulation of friable, loose loess on the divides
and fine-grained alluvial sediments along the valleys. Around
the hilly Sierras, a skirt of loess and piedmont deposits is
traversed by massive, brownish, clayey silts, with subordinate
sand content, layered in 2-3 m thick beds usually delimited
by paleosols.

The material is essentially wind-blown volcanic ash and dust,
but silt- or sand-sized clay pellets are also common. Mean
grain-size increases towards the west and the southwest.

The best known area in terms of the continental Holocene is
Sierra de la Ventana (Table 1). Recent works by Rabassa et al.
(1985a) and Rabassa (1985) have shown that loess accumulation
has been almost continuous during the entire Holocene epoch,
except for two major pedogenetic episodes which took place
(i) at the local Pleistocene-Holocene conventional boundary
(not necessarily at 10,000 BP) and (ii) sometime between
2000 yr BP and 1000 yr BP.

Finally, a 1.5 m thick sandy loess, the Matadero Saldungaray
Fm., was deposited probably immediately before and during the
"Little Ice Age" glacial readvance in the Patagonian Andes
(XVII th. to XIX th. centuries; Rabassa et al., 1984a; see
1.4).

The age of this unit has been established by limiting
(maximum) radiocarbon dates, palaeontological and
archaeological remains (Rabassa et al. 1985b; Salemme et al.
1985).

The beginning of the Holocene of Buenos Aires Pampas is
characterized by the extinction of **megamammals**, i.e., the
transition between the Lujanian Land Mammal Age and the
present land mammal associations (Pascual et al., 1966; Pascual
1984). However, the extinction of the Pleistocene mammals was
gradual and did not happen synchronously over the entire area.

In fact, survivors of these megamammals passed such
boundary and their remains have been found in intimate
association with Man (Politis 1984a; Fidalgo et al. 1986), at
ages as young as 6000 14-C yr BP.

The Holocene climate of this region has been estimated as
"wet-warm" by Fidalgo (1979) and Tonni and Fidalgo (1978),
based on geological and palaeontological criteria.
Palynological evidence (D'Antoni et al. 1985, Nieto et al.
1985, Fernandez and Romero 1984) suggests that the increase
of humidity and temperature was reversed towards the end of
the Flandrian transgression (Las Escobas Fm.; i.e., 4000 14-C

Table 2. Holocene of Northeastern Argentina (Iriondo, 1984).

AGE	LAND-MAMMAL AGE	PARANA RIVER	WESTERN ALLUVIAL FANS	BRAZILIAN SHIELD	PAMPEAN REGION	CLIMATE
HOLOCENE	PRESENT	PRESENT FLOOD PLAIN AND DELTA	SAN GUILLERMO Fm (AEOLIAN)	FLOOD PLAINS AND TERRACES OF MINOR RIVERS	SAN GUILLERMO Fm (AEOLIAN)	PRESENT CLIMATE / ARID
		LA PICADA Fm				
UPPER PLEISTOCENE	LUJANIAN	SALADILLO DEPOSITS	CAÑADA DE LAS VIBORAS Fm	HERNAN-DARIAS Fm / YUPOI Fm	BONAERENSE (PAMPEAN) DEPOSITS	SEMIARID COLD / HUMID

yr bp), leading to drier, steppe-like conditions, until present times, with minor fluctuations. Finally, a very short but dry episode occurred during the XVIII th century and more precisely, between AD 1698 and AD 1791 (Politis 1984b), supporting the hypothesis of humidity decrease during "Little Ice Age" times (Rabassa **et al.** 1985b).

2 THE HOLOCENE OF THE MESOPOTAMIAN AND CHACO REGION

The northeastern portion of Argentina consists almost completely of a wide plain composed of Quaternary sediments of alluvial, aeolian and marshy origin (Iriondo 1984). The regional slopes are very low and the local relief is almost unexistant.

The climate is presently very warm and wet, sharing most characteristics of neighbouring tropical and subtropical areas.

The geological evolution of this area during the Quaternary is defined upon the characteristics and spatial relationships of four major environments: the Río Paraná, the western alluvial fans, the Brazilian Shield and the Pampean Region. These environments have survived in the area throughout the Holocene (Table 2).

Neotectonics had a significant influence in this region, vertical movements have been interpreted to be very gentle but with clear surficial expression due to the shallow relief. This region is composed of tectonically tilted blocks and the youngest movements are probably modern displacements of older structures (Iriondo 1984).

The Pleistocene history of the region has been controlled by the changes in position and regime of the Río Paraná. These conditions, together with the climatic changes, prevailed during the Holocene as well.

The Early Holocene is represented by alluvial deposits of the La Picada Fm. along the fluvial valleys of southern Mesopotamia, thus suggesting a wetter climate. The Late Pleistocene loess was partially eroded in some areas, originating a micro-relief on the order of 1 m. During the Late Holocene, between 3,000 and 1,000 14-C yr BP (Iriondo 1981), an arid climate was established and the San Guillermo Fm., loess and aeolian sands, was deposited (Table 2). The present climate is wet subtropical in most of the area and it favours fluvial dynamics and soil genesis in the inter-fluvial areas (Iriondo 1984).

3 THE HOLOCENE OF THE CENTRAL ANDES

The Central Andes of Argentina and Chile is characterized by
dry, vegetation-less, very high mountain chains, which
develop an impassable barrier for the moisture bearing Pacific
winds. Therefore, precipitation falls mainly on the Chilean
side. On the mountains themselves, it is restricted to winter
snowfall that supports receding glaciers (Cobos and Boninsegna
1983).

The piedmont and lowland areas that extend east of these
chains are in fact very warm deserts, crossed by the outflowing
glacial streams. These rivers have been used since the XVI th
century to generate several fertile agricultural oases, the
city of Mendoza being the most important.

These mountain belts are composed of many different rock
types, from Early Cambrian marine beds to Late Cenozoic
volcanics. Many tectonic episodes have contributed to the
upbuilding of these ranges, but the Cenozoic Andean Tectonics
are responsible for their present distribution and
configurations.

The Andean Tectonics are still active. Several Neotectonic
events took place during the Quaternary and even a slight
uplift has been detected in Holocene times (Polanski 1963;
Fernandez 1984).

The Holocene of the Mendoza region has been characterized
by an episode of fluvial erosion and degradation in the
mountainous areas, together with generalized glacial recession
since the end of the Pleistocene, Holocene rock glaciers have
formed under these circumstances (Barsch and Happoldt 1985).

In the piedmont areas, loess-like silts and fine aeolian
sands (El Zampal Fm.) increase their thickness away form the
mountain front and cover the entire plain. The main source
for these sediments is volcanic ash, derived the result of
the "Postglacial Volcanic Association" (Polanski 1963), a
number of Holocene eruptions in the area.

Glacier fluctuations in the region of Río Atuel have been
established by Stingl and Garleff (1985), with glacier
expansion and up to 75% snowline depression (compared with
Late - Glacial Maximum) around 5,200 - 4,300 14-C yr BP, and
two much-reduced readvances during the "Little Ice Age"
(Garleff and Stingl 1985). No traces of a neoglacial expansion
around 2,200 - 2,000 yr BP have been detected yet by these
authors.

Dendrohydrological and glacier fluctuations studies in this

same area (Cobos and Boninsegna 1983) would indicate that the
"Little Ice Age" had already started around AD 1636. Annual
runoff data reconstructed back to AD 1575 suggest that periods
of high flow would correlate with glacial advances, around
AD 1636 - 1646, AD 1742 - 1754 and AD 1823 - 1850. The
filtered series shows a gradual diminishing tendency in the
annual river flow; after AD 1850 when irreversible glacier
recession started, the record keeps sustainedly below average.

Furthermore south, significant Holocene volcanic activity
has been identified by Brunotte (1985) around Laguna Llancanelo
(approx. lat. 35°S), recognizing middle Holocene ignimbrites
(PayúnMatrú Fm.) and Quizapú Volcano tephra (AD 1932). Many
undated Holocene volcanics are also present elsewhere into the
Northernmost Patagonian Andes.

4 THE HOLOCENE OF THE PATAGONIAN ANDES AND TIERRA DEL FUEGO

The Patagonian Andes are a N-S- trending range, extending
from lat. 38°S to Tierra del Fuego and the Cape Horn, that
block the fore entrance of the cool, wet westerlies into
the Patagonian Region. However, they are not high enough
to prevent the partial down-movement of the incessant storms
along the eastern slopes. Thus, the western coast is extremely
wet (above 7,000 mm/yr around lat. 50°S, Chile, Heusser, 1984)
whereas the eastern slopes show a significant decreasing rain
gradient which limits the extent of the Patagonian Forest just
to the E foot of the mountains.

The entire area was glaciated during the Pleistocene, in the
shape of a mountain ice - sheet that reached to the present
Pacific Ocean submarine platform south of lat. 43°S and,
across the Patagonian tablelands, to the Atlantic present
shoreline south of lat. 51°S (Caldenius, 1932; Holling and
Schilling, 1981).

The Pleistocene Fuego-Patagonian Ice - Sheet was reduced
during the Holocene to three isolated, much smaller remnants,
the Northern and Southern Patagonian Ice Caps and the Cordillera Darwin Ice Cap, in Chilean Tierra del Fuego. In spite
of that, they are still the larger ice bodies of the Southern
Hemisphere (outside of Antarctica), with an area of more than
26,000 sq. km.

Neotectonic activity should not be neglected. However, it
seems to be not as significant as in the Central Andes. It
should be considered, instead, the effect of the glacio-

Table 3. Tephrochronology of Patagonia.

(a) AUER (1956)

UNIT		AGE (yr.b.p.)
TEPHRA	IV	
TEPHRA	III	800 - 1,500
TEPHRA	II	2,300 - 4,000
TEPHRA	I UPPER	
TEPHRA	I LOWER	8,500 - 8,800
TEPHRA	O - BED O2	
	BED O	10,500 - 10,800
	BED O1	

(b) LAYA (1977) — RIO PIRECO FM.

	AGE (yr.b.p.)
ANCANTUCO MEMBER	
LAGO TOTORAL BED	A.D. 1960
BURIED ORGANIC SOIL	
EL RINCON BED	A.D. 1920
CORRENTOSO PALEOSOL	
LAGO ESPEJO MEMBER	
ANGOSTURA PALEOSOL	> 200
RIO PEREYRA MEMBER	1,599 ± 84
RIO BLANCO MEMBER	
PUYEHUE PALEOSOL	
LAGO MASCARDI MEMBER	
UPPER BED	
MIDDLE BED	
LOWER BED	

AB
CD

isostatic rebound after the vanishing of the Pleistocene ice
sheet, which started sometime around 14,000 years ago (Heusser
1983; Mercer 1976). This tectonic glacioisostatic uplift
reached perhaps a significant rate of more than 1 mm/yr
during most of the Holocene (and even the Late Holocene)
along the Beagle Channel (Urien, 1966; Porter **et al.** 1984;
Rabassa **et al.** 1986).

The Holocene of this region is dominated by two different
kinds of events: 1) the volcanic eruptions, mainly tephras,
from the many existing vents, some of them still active, and
2) the recent glacier fluctuations, generally named as Neoglaciations. The Patagonian tephras were widely studied by
Auer (1956), who based his observations upon the wrong
assumption that these pyroclastic eruptions had simultaneously
covered most of Patagonia.

He proposed a "tephrochronology" on the basis of five major
units (Table 3a). More recently, Laya (1977) re-established
a formal stratigraphy for these units in Northern Patagonia
(Table 3b). Finally, Crivelli Montero and Silveira (1983)
dated Auer's Tephra 1 Laya's Río Blanco Member-as "shortly
before 2,700 14-C yr BP ". The Neoglacial episodes have been
studied mainly by Mercer (1968, 1970, 1976, 1982, 1985;
Table 4). A regional summary has been presented by Fidalgo
and Rabassa (1984).

Recent unpublished work by Friedrich Rothlisberger (written
communication) has provided evidence of up to twelve Holocene
glacier readvances, peaking at 8600, 8200, 5200, 4600, 3300,
2700, 2200. 1300, 900, 600, 300 and 150 years BP . These data
are sustained on very extensive field work and a large amount
of radiocarbon dates (see also Rothlisberger and Geyh, 1985).

Mercer has suggested that the Wisconsinan glaciation named
as Llanquihue and Nahuel Huapi glaciations in Chile and
Argentina respectively (Flint and Fidalgo, 1964; Mercer, 1976),
had finished before 12,000 BP . Moreover, Mercer (1976) denied
the possibility of Late Glacial readvances in the Patagonian
glaciers about 11,000 BP , as it happened in Scandinavia
during the Younger Dryas stade.

This point of view is conflicting with Heusser (1974) and
Heusser and Streeter's (1980) data, who suggested that a
period of mean annual temperatures as low as during the end
of the Last Glaciation occurred between 11,300 and 9,400 yr
BP. The regional Hypsithermal would have taken place only
between 9,000 and 6,000 BP , approximately. Later, three
climatic events, cooler and wetter than the present climate,

Table 4. Holocene Glacier Fluctuations in Patagonia (Mercer, 1976).

NEOGLACIATION		AGE B.P.
3 b	"LITTLE ICE AGE"	0 - 500
3 a		600 - 900
2		2000 - 2700
I		4200 - 4750
HIPSITHERMAL		5500 - 10.000 ?
WISCONSIN GLACIATION		> 14.500

would have occurred in the area, around 4,950-3,160; 2,000 and 350 yr BP (Heusser and Streeter, 1980). These episodes would be roughly coincident with the "Neoglaciations" as described by Mercer (1976; see Table 4).

According to Mercer (1976), the First Neoglacial was the largest one; however, there are examples of glaciers which reached their maximum expansion sometime later. The Last Neoglaciation, called the "Little Ice Age" in the Northern Hemisphere, would have developed in two major fluctuations, between 500 - 350 yr BP and 250 - 0 yr BP, with a relatively warmer intermediate event (Grove, 1979).

The palynological studies of Markgraf (1980; see also Valencio et al. 1985) suggest that the **Nothofagus dombeyi** forest expanded as early as 13,000 yr BP in Northern Patagonia, reaching a maximum density around 11,500 yr BP. Later, around 8,500 yr BP the other **Nothofagus** genus would have been incorporated to the forest as well. This slow succession is interpreted by Markgraf (1980) as the result of mean annual temperatures not high enough to allow the forest expansion.

Later, increasing dryness periods would have occurred around 6,500 BP and 5,000-4,500 BP.

These data are contradictory with Heusser and Streeter's (1980) results; Markgraf (1980) explains it as a consequence of different environmental conditions on both sides of the Andes, thus oversteepening the present precipitation gradient yet even more.

"Little Ice Age" readvances have been studied in detail in Northern Patagonia by Rabassa **et al.** (1984a; 1985c).

The results of the dendrochronological and lichenometric analyses of "Little Ice Age" moraines are presented in Table 5.

This glacial fluctuation took place between the XVII and the XIX th. centuries. It probably reached a maximum in this region at the end of the XVIII th. century and it was over in a few decades after AD 1850. The climatic correlation of this glacier expansion and the aridization of the western Pampas has also been proposed (Rabassa **et al.** 1985b; Politis 1984b). The suggested chronology is in rough agreement with similar advances in other parts of South America, the Northern Hemisphere and New Zealand (Grove 1979; Garleff and Stingl 1985; Rothlisberger and Geyh, 1985).

In southernmost South America and Tierra del Fuego, palynological studies by Heusser (1984) suggest that an early expansion of **Nothofagus** forest took place before 12,730 yr BP. A colder and drier episode, with the spread of tundra, occurred between 12,730 and 10,080 yr BP. Conditions became warmer and wetter, with extension of closed **Nothofagus** forest larger than today until 5,520 yr BP. After then, climate would have turned rather colder and drier again, probably coinciding with the Neoglaciations' period.

The insular condition of Tierra del Fuego, with the final opening of the Magellan Strait and the Beagle Channel, would have been completed around 7,900 yr BP (Porter **et al.**, 1984; Rabassa **et al.** 1986).

CONCLUSIONS

Much work has yet to be done concerning the Holocene of Argentina. The most significant problems still remain the lack of fine stratigraphical work, radiocarbon dates and palynological analyses in certain areas, that should enable us to improve the preliminary regional correlations attempted in this paper and to build a general chronostratigraphic framework for the entire country (Table 6).

Table 5. Little Ice-Age and later advances in Mount Tronador and Volcán Lanin (Rabassa et al., 1984a).

MATURE (PRE-GLACIAL) FOREST	RIO MANSO GLACIER AD 1656	CASTAÑO OVERO GLACIER AD 1629		LANIN N SLOPE > AD 1742
M 1	AD 1772		?	GROUP "A" 1746-1788
M 2	AD 1812	M 1/2	AD 1818-1829	GROUP "B" 1815-1839
OUTER M 3	AD 1853	M 3	AD 1842	
INNER M 3	AD 1860	M 4/M5	AD 1857	
M 4 a	AD 1896	M 6	AD 1884	GROUP "C" 1851-1896
M 4 b	AD 1902	M 7	AD 1902	
M 5	AD 1923		?	GROUP "D" 1938-1957

✳ From RABASSA et al 1984 a

284

Table 6. Chronological, climatic and faunistic correlation of the late Pleistocene and Holocene of Argentina.

				YR BP
	HISTORICAL TIMES	LITTLE ICE AGE	INTRODUCTION OF EUROPEAN FAUNA	500
THE HOLOCENE	LATE HOLOCENE	NEOGLACIATIONS COOLER - DRIER	RETRACTION OF BRAZILIAN FAUNA	4500
	MIDDLE HOLOCENE	HIPSITHERMAL TRANSGRESSION WARMER - WETTER	EXPANSION OF BRAZILIAN FAUNA	6000
	EARLY HOLOCENE	GRADUAL CLIMATE AMELIORATION COOLER - DRIER	ARID - ENVIRONMENTS EXTANT FAUNA-LAST SPECIMENS OF EXTINCT FAUNA	10000
THE PLEISTOCENE	LATE PLEISTOCENE	LATE - GLACIAL	BEGINNING OF MASSIVE EXTINCTIONS	15000
		NAHUEL HUAPI GLACIATION VERY COLD VERY DRY	PLEISTOCENE FAUNA (LUJANENSE MAMMAL - AGE)	

The spread of the interest on the Holocene and the appearance and growth of new research groups all over Argentina will certainly let us achieve these goals in the near future.

REFERENCES

Ameghino, F. 1889. Contribución al conocimiento de los mamíferos fósiles de la República Argentina. **Act. Acad. Nac. Cienc. Córdoba**, 6:1-1027. Córdoba.
Auer, V. 1956. The Pleistocene of Fuego-Patagonia. Part I. The ice and interglacial ages. **Annales Academinae Scientiaum Fennicae**, III, Geologica Geographica, 45:1-226. Helsinki.

Barsch, D. and Happoldt, H. 1985. Blockgletscherbildund und holozäne Höhenstupengliederung in den mendozinischen Anden, Argentinien. **Zbl. Geol. Paläont**, Teil I, **HF.11/12**, p.1625-1632. Stuttgart.

Brunotte, E. 1985. Reliefent wicklung am Westrand des Beckens der Laguna Llancanelo, Argentinien. **Zbl. Geol. Paläont.** Teil I, **Hf 11/12** p.1571-1580. Stuttgart.

Caldenius, C. 1932. Las glaciaciones cuaternarias en la Patagonia y Tierra del Fuego. **Geografiska Annaler**, 14:1-164.

Cobos, D. and Boninsegna, J. 1983. Fluctuations of some glaciers in the Upper Atuel River basin, Mendoza, Argentina. **Quaternary of South America and Antarctic Peninsula**, 1:61-82. A.A. Balkema, Rotterdam.

Crivelli Montero, E. and Silveira, M. 1983. Radiocarbon chronology of a tephra layer in Río Traful Valley, Province of Neuquén, Argentina. **Quaternary of South America and Antarctic Peninsula**, 1:135-150. A.A. Balkema, Rotterdam.

D'Antoni, H.; Nieto, A. and Mancini, M. 1985. Pollen analytic stratigraphy of Arroyo Las Brusquitas Profile (Buenos Aires Province, Argentina). **Zbl. Geol. Paläont.**, Teil I, **Hf. 11/12**, p.1721-1730. Stuttgart.

Fernandez, C. and Romero, E. 1984. Palynology of Quaternary Sediments of Lake Chascomús, Northeastern Buenos Aires Province, Argentina. **Quaternary of South America and Antarctic Peninsula**, 3:201-221. A.A. Balkema, Rotterdam.

Fernandez, B. 1984. Stratigraphy of the Quaternary piedmont deposits of the Río de las Tunas Valley, Mendoza, Argentina. **Quaternary of South America and Antarctic Peninsula**, 2:31-40. A.A. Balkema, Rotterdam.

Fidalgo, F. 1979. Upper Pleistocene - Recent Marine Deposits in Northeastern Buenos Aires Province (Argentina). **In: Proceedings 1978 Int. Symp. Coastal Evolution in the Quaternary**, K. Suguío, et al, eds., São Paulo, Brasil, p.384-404.

Fidalgo, F. and Tonni, E. 1982. The Holocene in Argentina, South America. **In: "Chronostratigraphical Subdivision of the Holocene"**, Striae, 16:49-52. Uppsala.

Fidalgo, F. and Rabassa, J. 1984. Los depósitos cuaternarios. **In: IX Congreso Geológico Argentino, 1984, Relatorio I (11):** 301-316. Buenos Aires.

Fidalgo, F.; Mec Guzman, L.; Politis, G.; Tonni, E. and Salemme, M. 1986. Investigaciones Arqueológicas en el Sitio 2 de Arroyo Seco (Partido de Tres Arroyos, Provincia de

Buenos Aires, Argentina). In: A. Bryan, ed., **"New evidence for the Pleistocene Peopling of the Americas"**, 380pp., Peopling of the Americas Publications, Orono, Maine.

Flint, R.F. and Fidalgo, F. 1964. Glacial geology of the border zone of the Andes between lat. 39° 10'S and 41° 20'S. Geol. Soc. Am. Bull., 75:335-352.

Garleff, K. and Stingl, H. 1985. Jungquartäre Klimageschichte und ihre Indikatoren in Südamerika. **Zbl. Geol. Paläont. Teil I, Hf. 11/12,** p.1769-1776. Stuttgart.

Grove, J. 1979. The glacial history of the Holocene. **Progress in Physical Geography,** 3(1):1-54.

Heusser, C. 1974. Vegetation and climate of the Southern Chilean Lake District during and since the last interglaciation. **Quaternary Research,** 4:290-315.

Heusser, C. and Streeter, S. 1980. A temperature and precipitation record of the past 16,000 years in Southern Chile. **Science,** 210:1345-1347.

Heusser, C. 1983. Quaternary palynology of Chile, **Quaternary of South America and Antarctic Peninsula,** 1:5-22. A.A. Balkema, Rotterdam.

Heusser, C. 1984. Late Quaternary climates of Chile. **In:** J.C. Vogel (ed.), Proceed. Int. Symp. SASQUA, **"Late Cainozoic Palaeoclimates of the Southern Hemisphere",** p.59-83, A.A. Balkema, Rotterdam.

Holling, J. and Schilling, D. 1981. Late Wisconsin-Weichselian mountain glaciers and small ice caps. **In:** G. Denton and T. Hughes (eds.), **"The Last Great Ice Sheets",** Wiley, New York, 484pp., p.179-206.

Iriondo, M. 1984. The Quaternary of Northeastern Argentina. **Quaternary of South America and Antarctic Peninsula,** 2:51-78, A.A. Balkema, Rotterdam.

Laya, H. 1977. Edafogénesis y paleosuelos de la formación téfrica Río Pireco (Holoceno), suroeste de la Provicia de Neuquén, Argentina. **Asociación Geológica Argentina Rev.,** 32(1):3-23. Buenos Aires.

Markgraf, V. 1980. Paleoclimatic reconstruction of the last 15,000 years in Subantarctic and temperate regions of Argentine. **XV Symp. Palynol. Langue Francaise,** París.

Mercer, J. 1968. Variations of some Patagonian Glaciers Since the Late Glacial. **American J. Science,** 266:91-109.

Mercer, J. 1970. Variation of some Patagonian Glaciers since the Late Glacial: II. **American J. Science,** 269:1-25.

Mercer, J. 1976. Glacial history of southernmost south America. **Quaternary Research,** 6:125-160.

Mercer, J. 1982. Holocene glacier variations in southern south
America. In: W. Karlén (ed.), Holocene Glaciers, **Striae,**
18:35-40.

Mercer, J. 1985. Changes in the Ice Cover of Temperate and
Tropical South America during the last 25,000 years. **Zbl.
Geol. Paläont,** Teil I, **Hf. 11/12,** p.1661-1665. Stuttgart.

Nieto, A. and D'Antoni, H. 1985. Pollen analysis of sediments
of the Atlantic Shore at Mar Chiquita (Buenos Aires Province,
Argentina) **Zbl. Geol. Paläont,** Teil I, **Hf. 11/12,** p.1731-1738.
Stuttgart.

Pascual, R.; Ortega Hinojosa, E.; Gondar, D. and Tonni, E.
1966. II. Las edades del Cenozoico mamalífero de la Provincia
de Buenos Aires. In: **Paleontografía Bonaerense,** A. Borrello
(ed.), IV, Vertebrata, p.3-27, C.I.C., La Plata.

Pascual, R. 1984. Late Tertiary mammals of southern South
America as indicators of climatic deterioration. **Quaternary
of South America and Antarctic Peninsula,** 2:1-30, A.A.
Balkema, Rotterdam.

Pasotti, P. 1974. La Neotectónica en la llanura pampeana.
Fundamentos para el mapa neotectónico. **Instituto de Fisio-
grafía y Geología, Publicación N° 58,** Universidad Nacional
de Rosario, Rosario.

Polanski, J. 1963. Estratigrafía, Neotectónica y Geomorfología
del Pleistoceno pedemontano, entre los ríos Diamante y Mendoza. **Asoc. Geol. Arg. Rev.,** 17(3-4):127-349. Buenos Aires.

Politis, G. 1984a. Arqueología del área interserrana bonaerense. **Tesis Doctoral,** Facultad de Ciencias Naturales y Museo,
Universidad Nacional de La Plata, unpublished.

Politis, G. 1984b. Climatic variations during historical times
in Eastern Buenos Aires Pampas, Argentina. **Quaternary of
South America and Antarctic Peninsula,** 2:133-162, A.A.
Balkema, Rotterdam.

Porter, S.; Stuiver, M. and Heusser, C. 1984. Holocene Sea
Level changes along the Strait of Magellan and Beagle
Channel, Southernmost South America. **Quaternary Research,**
22:59-67.

Rabassa, J.; Brandani, A.; Boninsegna, J. and Cobos, D. 1984.
Cronología de la "Pequeña Edad del Hielo" en los glaciares
Río Manso y Castaño Overo, Cerro Tronador, Provincia de
Río Negro. **IX Congreso Geológico Argentino, Actas,** 3:624-639.
Buenos Aires.

Rabassa, J. 1985. Geología de los depósitos del Pleistoceno
Superior y Holoceno en la cabecera del Río Sauce Grande.

I Jorn. Geol. **Bonaerenses**, Abstracts, Tandil. Full paper
in press.

Rabassa, J.; Brandani, A.; Salemme, M. and Politis, G. 1985a.
La "Pequeña Edad del Hielo" (Siglos XVI a XIX) y su posible
influencia en la aridización de áreas marginales de la Pampa
Húmeda (Provincia de Buenos Aires). I **Jorn. Geol. Bonaeren-**
ses, Abstracts, Tandil. Full paper in press.

Rabassa, J.; Heusser, C.; Salemme, M.; Politis, G. and
Stuckenrath. R. 1985b. Hallazgo de troncos fósiles (**Salix**
humboldtiana) en sedimentos aluviales del Río Sauce Grande,
Provincia de Buenos Aires. I **Jorn. Geol. Bonaerenses**,
Abstracts, Tandil. Full paper in press, **Ameghiniana**, Buenos
Aires.

Rabassa, J.; Brandani, A.; Boninsegna, J. and Cobos, D. 1985c.
Glacier fluctuations during and since the Little Ice Age and
forest colonization. Monte Tronador and Volcán Lanín,
Northern Patagonian Andes, Argentina. **In: International**
Symp. on Glacier Mass balance, Fluctuation and Runoff, Alma
Ata, USSR, Abstracts, p.86-87, Full paper in press.

Rabassa, J.; Heusser, C. and Stuckenrath, R. 1986. New data
on Holocene sea transgression in the Beagle Channel, Tierra
del Fuego. Int. Symp. on Sea-Level changes, São Paulo,
Brazil, July 1986. **Quaternary of South America and Antarctic**
Peninsula, 4:291-310, A.A. Balkema, Rotterdam.

Rothlinsberger, F. and Geyh, M. 1985. Gletscherschwankunger
der Nacheiszeit in der Cordillera Blanca (Peru) und den
südlichen Andes Chiles und Argentiniens. **Zbl. Geol. Paläont.,**
Teil I, H. 11/12, p.1611-1614. Stuttgart.

Salemme, M.V.; Politis, G.; Madrid, P.; Oliva, F. and Güerci,
L. 1985. Informe preliminar sobre las investigaciones arqueológicas en el Sitio La Toma, Partido de Coronel Pringles
(Provincia de Buenos Aires). **VIII Congreso Nacional Arqueo-**
logía Argentina, Concordia, Abstracts, p.3. Full paper in
press.

Stingl, H. and Garleff, K. 1985. Spätglaziale und holozäne
Gletschekund Klimaschwankungen in den argentinischen Anden.
Zbl. Geol. Paläont. Teil I, Hf. 11/12, p.1667-1678.
Stuttgart.

Tonri, E. and Fidalgo, F. 1978. Consideraciones sobre los
cambios climáticos durante el Pleistoceno tardío-Reciente
en la provincia de Buenos Aires. Aspectos ecológicos y zoc-
geográficos relacionados. **Ameghiniana**, 15(1-2):235-253.
Buenos Aires.

Valencio, D.; Creer, K.M.; Sinito, A.M.; Mazzoni, M.; Alonso,
    M.S. and Markgraf, V. 1985. Palaeomagnetism, sedimentology,
    radiocarbon age determinations and palynology of the Llao-
    Llao area, southwestern Argentina (lat. 41°S; long. 71°30'W):
    Palaeolimnological aspects.
    Quaternary of South America and Antarctic Peninsula, 3:109-
    148, A.A. Balkema, Rotterdam.
Urien, C. 1968. Edad de algunas playas elevadas en la Penín-
    sula de Ushuaia y su relación con el ascenso costero, post-
    glaciario. **Actas III Jorn. Geol. Arg.**, 2:35-41, Buenos Aires.

Shoreline changes in French Guiana

ABSTRACT

Shoreline changes in French Guiana are discussed within three
temporal scales: present-day dynamic variations, Holocene
shoreline changes and Late Quaternary sea-level variations.
Some remarks concerning sea-level data are proposed in
conclusion.

RESUMEN

Los cambios en la línea de costa en la Guayana francesa se
discuten dentro de tres escalas en el tiempo: las variaciones
dinámicas actuales, los cambios en la línea de costa durante
el Holoceno y las variaciones del nivel del mar durante el
Cuaternario Tardío. En conclusión, se proponen algunas observaciones con respecto a datos sobre el nivel del mar.

INTRODUCTION

French Guiana coastal evolution involves some important
problems concerning sea-level research of the north of South
America.

This paper deals with shoreline changes within three temporal
scales: present-day dynamic variations, Holocene shoreline
changes and Late Quaternary sea-level variations.

Present-day shoreline changes in French Guiana - and within
the Guiana's region - are striking and specific, directly
linked with the huge Amazon discharge. Fine-grained sediments

are transported from the Amazon's mouth to the Orinoco delta
by the action of currents and swell. One part of this immense
supply moves like extensive shoreface-attached mudbanks, there
are continually migrating to the W-NW. At the same time
shoreline is prograding and retreating.

Field indicators of Holocene shoreline changes are cheniers,
narrow relic beach ridges which crossed, sub-coastal paralic
environment. They represent relic beach ridges.

An evolutionary morpho-sedimentary model of the shelf during
the past 30.000 years is proposed by I.G.B.A. researchers. The
part of glacio-eustatic changes is stressed.

Some remarks concerning sea-level data are proposed in
conclusion.

1 PRESENT DAY SHORELINE CHANGES

Present-day shoreline changes in French Guiana are directly
linked with the huge sediment discharge of the Amazon River:
approximately 250 million m^3 of fine-grained sediments are
transported annually along the Guiana's coastline by the
equatorial current and longshore currents created by trade-
wind-driven waves (Wells and Coleman, 1977; Rine and Ginsburg,
1985). Evidence for this northwest transport is supplied by
Nedeco (1968), Allersma (1971), Eisma and Van der Marel (1971)
and Gibbs (1976).

One part of Amazon's output moves in suspension; the other
part (some 110 million m^3/y) moves in the form of very large
shoreface-attached mudbanks which migrate continually westwardly. Rine and Ginsburg (1985) defined a typical mudbank
in Suriname as up to 5 m high, 50 to 60 km long, 10 to 20 km
wide, and oriented at oblique angles to the coastline. In
French Guiana there are presently 6 mudbanks with
approximately 20 to 40 km long (Froidefond J.M. **et al.**, 1985).

When a mudbank is attached to the coast, the shoreline is
undergoing progradation. Between two mudbanks - within the
interbank zone - shoreline is undergoing erosion and is
retreating. The shoreline is constantly changing because this
dynamic creates very short-term morphological variations.
Measurements have been taken of the migration's rate of
mudbanks and interbank zones along the Guiana's coast. In
French Guiana recent research (CORDET-IGBA Report, 1985)
showed that between 1979 and 1984 mudbanks have moved in a

W/NW direction at speeds varying from 250 and 1250 m/y. The
average rate of migration is 900 m/y (Froidefond J.M. **et al.**,
1985; Froidefond J.M. **et al.**, in press).

Frequency and intensity of sedimentation varies along the
coast. Within the mudbank zones sedimentation is high because
muds are generally so fluid that they interact with surface
waves causing them to be altered and damped (Rine and Ginsburg,
1985). Within the interbank zones, on the contrary, wave
energy is relatively high because of lack of wave dampening,
producing erosion of the shoreline and shoreface.
Sedimentation's rate varies also during the year: long term
periods of higher wind speed (December to May; Nedeco, 1968;
Allersma, 1971) seems to be important and coincide with
periods of high concentrations of suspended matter within the
coastal waters (Rine and Ginsburg, 1985). In fact, during this
annual period of high winds we observed, within an interbank
zone of the west coast (Pointe Isére), high-waves amplitudes,
washover process and heavy erosion.

In conclusion:

- The most striking phenomenon influencing the present-day
coastal evolution is the supply of the enormous mass of fine-
grained sediments discharged by the Amazon River and carried
steadily westward by the combined action of swell and currents.
- Mudbanks and interbank zones move westwardly along the French
Guiana shoreface. Evidence of this northwest transport is also
supplied by the deflected mouths of several rivers (Sinnamary,
Iracoubo, Mana...).
- Rine and Ginsburg (1985) showed how the stacking of
sediments from migrant mudbanks creates on the shoreface a
vertical sequence of laminated and massive muds with
discontinuity features and, on the coastal plain, a dynamic
horizontal sequence of mud marshes and sand cheniers.

2 HOLOCENE SHORELINE VARIATIONS

French Guiana Holocene coastal plain is roughly situated
between 0 and 5 m high and is composed of marine clays
(Demerara Formation), waterfront mangrove and swampland. The
latter is criss-crossed by narrow cheniers that are shallow -
based perched sandy ridges which rest on clay.

Cheniers have a general disposition roughly parallel to the

present shoreline. Two major systems of cheniers have been recognized in French Guiana owing to their morphological situation and to their pedological evolution. The "old" cheniers are generally situated between the marine clays of the Mara Phase (8000/6000 BP) and those of the Moleson Phase (2600/1300 BP). "Recent" cheniers are of the Moleson Phase (2600/1300 BP). "Recent" cheniers are separating the Moleson clays from those of the Comowine Phase (1000 BP to Present).

We observed, however, that cheniers disposition could be much more complicated, particularly within the estuaries environment. On the west coast, for instance, cheniers disposition are rather complex in character (Prost M.T., 1987). Furthermore, in the present-day cheniers formation continues to occur in the interbank zones (for instance, along a part of the west coast). Cheniers' formation takes place in a narrow zone around high water level; ridges migrate westwardly by beachdrifting. When more sand is being removed than deposited, cheniers are eroded, particularly during highs winds period and high-waves amplitudes.

The chenier's sands have local origine (Lafond L.R., 1967; Krook L., 1968; Turenne J.F., 1978; Froidefond J.M. and **al.**, 1985). The analysis of grain-size distribution have been done in the ORSTOM-Center (Cayenne) as well as preliminary remarks on mineralogical composition and on shape, roundness and surface texture of quartz grains of "raw" samples. Sandy formations are roughly similar as well as in present-day beaches and in cheniers. Heavy mineral composition should be defined to supply further information, but there is no doubt that cheniers represent relic beach ridges.

Rine and Ginsburg (1985) claimed that chenier plain morphology in Suriname was created by lateral stacking of mudbank deposits (marshes) separated by interbank beaches (cheniers). Migration of mudbanks is probably "a process that has occurred annually for the last 3.500 years" (Rine, 1980). If we accepted Rine's conclusion that means that vertical changes on sea-level during Holocene are only one componenet among others factors which influence French Guiana's shoreline evolution.

3 LATE QUATERNARY SEA-LEVEL CHANGES

An evolutionary morpho-sedimentary model of the French Guiana shelf during the Late Quaternary is proposed by Bouysse **et al.** (1977); Jeantet D. (1982) and Pujos and Odin (1986). It implies major sea-level changes and accounts for two trangressions and one regression. The Amazon River controls the sedimentary pattern of the shelf in relation with glacio-eustatic sea-level changes.

During a high sea-level (30.000 BP) the suspended load of the Amazon was discharged by the Guiana's current over the inner shelf in the form of a massive mud wedge. This "fossile" mud is presently below the present mud wedge. As the fossile and the modern mud present similar characteristics Pujos and Odin think that paleogeographical environment's conditions were similar to those of present-day.

During low glacio-eustatic sea level stillstands (-100/-90 m) the process was modified. The shoreline regressed to the shelf edge and the Amazon sediments were directly channeled onto the abyssal plain via the Amazon cone and deep sea fan. On the platform there are remains of a barrier reef (17.000/ 12.000 BP) and relict facies (rubified coarse-grained sands of the Maroni's delta, fine to very fine "verdine" sands, etc). When the sea-level was near present 30 m isobath, the suspended load of the Amazon begins to accumulate again, at about 8.000 BP (Pujos and Odin, 1986).

This evolutionary model shows how the epicontinental sedimentation took place on the French Guiana stable margin adjacent to an old peneplain and stress the Amazon River's controls on the coastline evolution. Sedimentation was continuous: fine-grained sediments from Amazon accumulate during high sea-levels and autochtonous sands and reef complexes develop during low sea-levels.

Remarks concerning paleoclimatic variations correlated with sea-level changes are done by Pujos and Odin (1986). The authors stress that between 20.000 and 17.000 BP climatic conditions were drier. In French Guiana "there are more climatic contrasts with sudden worsening of drought conditions", thereby allowing the extension of savanna to the detriment of the forest. That coincides with "intense evacuation of continentally-derived sediments". The region which is at present overlain by seawater masses is transected by river valleys and locally occupied by lagoons. From about 12.000 BP to the present, during sea-level rising, the climate becomes gradually similar to the present and there are extension of forests and mangroves at the expense of savannas (Pujos and Odin, 1986).

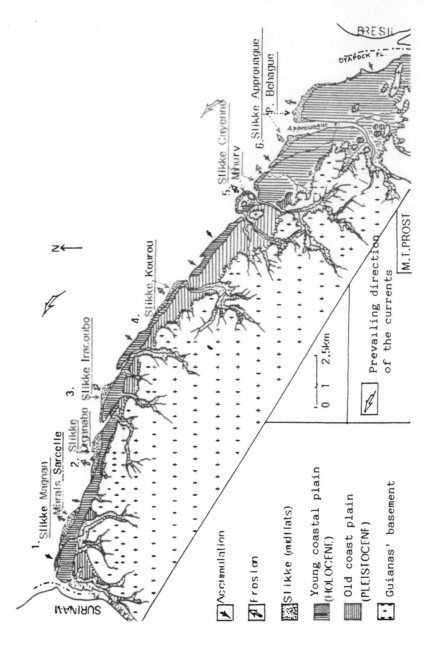

Figure 1. French Guiana Coastal Plain, July 1986. Prograding and retreating area.

Legend:
- Accumulation
- Erosion
- Slikke (mudflats)
- Young coastal plain (HOLOCENE)
- Old coast plain (PLEISTOCENE)
- Guianas' basement

1. Slikke Magnan
   Marais Sarcelle
2. Slikke Organabo
3. Slikke Iracoubo
4. Slikke Kourou
5. Slikke Cayenne
   Mahury
6. Slikke Approuague
   P. Behague

SURINAM

BRESIL

OYAPOCK FL.

Approuague

N

0 1 2,5km

Prevailing direction of the currents

M.T.PROST

4 DISCUSSION

Shoreline changes in French Guiana involves several problems
concerning sea-level research. Studies of the Holocene coastal
plain and of the continental shelf, based on field records,
showed a variety of shoreline changes indicators.

On one hand, the morpho-sedimentary model of the French
Guiana shelf evolution from the Late Quaternary to 8.000 BP
draws attention to glacio-eustatic sea-level changes. The
variations on the Amazon River's supply are connected with
these changes. During Holocene, after the maximum sea-level
rise of 6.000 BP, Brinkman and Pons (1968) indicate a "stable"
sea-level. But relative sea-level changes in Suriname coast
must be regarded with caution because until now the part of
relative changes in land level lacks of definition.

On the other hand, Rine and Ginsburg (1985) demonstrated
that chenier coastal plain of Suriname was formed by lateral
stacking of mudbanks deposits separated by interbank beaches,
dynamic that has probably ocurred since 3.500 BP. Moreover,
present-day coastal evolution clearly shows that shoreline
changes are directly linked-up with the sedimentary input
of the Amazon River and with coastal hydro-dynamic conditions.

Discriminate the nature of sea-level changes in French
Guiana is still extremely difficult. We need much more
research on paleogeographical and paleoecological environment
and on present-day coastal area evolution. Only detailed and
rigourous studies within the regional context may provide
new SL data and reassessment of existing studies. On one hand,
IGCP Projects 61 and 200 have shown that no part of the earth's
crust can be considered stable and that the relationship
between eustatic change and isostatic movements is very
complex. On the other hand, the concept of geoidal eustasy
shows that eustasy as a world-wide phenomenon "can no longer
be regarded as valid" (O. Von Plassche, 1986). Curves based
in detailed studies (eg. Brinkman and Pons, 1968) but without
considering geoidal variations must be referred to with very
high caution. In the Brazilian coast, on the contrary, high
sea-levels are interpreted as due to variations in the geoid
surface (Martin **et al.**, 1979-1980). Furthermore, Newman W.S.
(1985) demonstrate, owing to an elevation vs. time plot of
more than 4000 radiocarbon-dated sea level indicators for the
past 16.000 years, that the magnitude of "eustatic" sea-level
rise is exceed by both neotectonic as well as geoidal changes
in level.

Morner A.N. (1986) stress that dynamic sea level changes
(linked with various meteorological, hydrological and
oceanographic factors) may occur very rapidly (rates in the
order of some 100 mm/y). Changes in oceanic circulation
system may induced dynamic sea-level changes of a wide,
geographical extension (Morner, 1986). They also lead to a
redistribution of the heat stored in the oceans, and may have
a climatic effect.

REFERENCES

Allersma, E. (1971). Mud on the oceanic shelf of Guiana **Symp. on Investigation and Resources of the Caribbean Sea and Adjacent Regions.** UNESCO. p.193-203. Paris.

Bouysse Ph.., Kudrass, H.R., Le Lann, F. (1977). Reconnaissance sédimentologique du plateau continental de la Guyane Francaise (mission Guyamer, 1975). **Bull. B.R.G.M IV, 2:141-179.**

Eisma, D. and Van der Marel, H.W. (1971). Marine muds along the Guiana Coast and their origin from the Amazon basin. **Contrib. of Mon: and Petr., 31:321-334.** Spring Verlag.

Froidefond, J.M., Prost, M.T. and Gribculard, R. (1985). Etude de I'évolution morpho-sédimentaire des littoraux argileux sous climat équatorial: l'exemple du littoral guyanais.

Rapport CORDET - I.G.B.A, 189pp. Dép. de Géol. et Océanographie. Univ. Bordeaux-I.

Froidefond, J.M., Pujos, M. & Andre, X. (in press). Measurements of the displacement of mudbanks between 1979 1984 along the coastline of French Guiana.

Gibbs, R.J. (1976). Amazon River sediment transport in the Atlantic Ocean. **Geology,** January, p.45-48.

Jeantet, D. (1982). Processus sédimentaires et évolution du plateau guyanais au cours du Quaternaire Terminal. **These 3eme cycle. Dép. Géol. et Océanographie I.G.B.A.** Univ. de Bordeaux - I. Talence.

Krock, L. (1986). Sediment petrographical studies in northern Suriname. **Thesis. Academish Proefscherift.** Vrije Univ. Amsterdam.

Lafond, L.R. (1967). Etudes littorales et estuariennes en zone intertropicale humide. **These es-Sciences.** 3 tomes. Paris.

Nedeco (1968). Surinam Transport Study. **Report on Hydraulic Investigation.** 23pp. The Hague.

Newman, W.S. (1985). Palaeogeodesy. In. **IGCP Project 200 Newsletter and annual report. January 1986.** Univ. of Durham. U.K.

Morner, N.A. (1984). Planetary, solar, atmospheric, hydrospheric and endogene processes as origine of climatic changes on the Earth. In, **N.A. Morner and W. Karlen Ed. Climatic changes on a yearly to millenial Basis.** Reidel Publ. Co., p.637-651.

Morner, N.A. (1985). Short-term paleoclimatic changes: observation data and causation mechanisms. In, **Proc. "A climatic symposium in hon. Prof. R.W. Fairbridge".** New York.

Morner, N.A. (1985). Geoidal research: general. In, **Newsletter and annual report IGCP 200 (January 1986).**

Morner, N.A. (1986). Eustasy, geoid changes and dynamic sea surface changes due to interchange of momentum. In, **Inter. Symp. of Sea-level Changes.** Qiongdao and Yantai. China. Ed. by Q.Y. Zhao Songling.

Prost, M.T. (1987). Les cotes des Guyanes: état des études. (Project PIGC-201). **Rapport interne ORSTOM.** Hydrologie-Géomorphologie. Centre ORSTOM. Cayenne. Guyane francaise.

Pujos, M. & Odin, G.S. (1986). La sédimentation au Quaternaire Terminal sur la plate-forme continentale de la Guyane française. **Oceanologica Acta,** 9(4):363-382.

Rine & Ginsburg (1985). Depositional facies of the mud shoreface in Suriname, South America. A mud analogue to sandy, shallow marine deposits. **Comparative Sed. Lab.** Univ. of Miami. p.633-651. Florida.

Rine, J.M. (1980). Depositional environments and Holocene reconstruction of an argillaceous mud belt. Surinam. South America. **PhD Dissert. Univ. of Miami.** 222pp.

Turenne, J.F. (1978). Sédimentologie des plaines cotieres. **Atlas de la Guyane** CNRS/ORSTOM.

Van de Plassche, O. Ed. (1986). Sea-Level Research: a manual for the collection and evaluation of data. **Geo-Book.** Norwich.

Wells, J.T. & Coleman, J.M. (1977). Nearshore suspended sediment variations, central Surinam coast. **Marine Geol.** v.24, p.M47-M54.

Wells, J.T. & Coleman, J.M. (1977). Longshore transport of mud by waves: northeastern coast of South America. **Geol. en Mijnb,** 57:353-359.

NORBERTO OLMIRO HORN FILHO
Center of Studies in Coastal and Marine Geology (CECO), Institute of
Geosciences (IG), Federal University of Rio Grande do Sul (UFRGS), Brazil

Geological mapping of the north coastal plain of Rio Grande do Sul, southern Brazil

ABSTRACT

The mapped area in this paper is situated in the northeast portion of the Rio Grande do Sul state, southern Brazil, between 29°15' and 29°45' south latitudes and 49°30' and 50°15' west longitudes, including the planialtimetric charts of Torres, Tres Cachoeiras, Arroio Teixeira and Maquiné, in 1:50.000 scale.

The main outcroping surface unities are the shallow marine, aeolian, lagoonal, deltaic, fluvial and alluvial fan sediments, that are simbolized respectively by Qbc, Qbd, Qp, Qd, Qf, and Ql. These sediments are seated on Gondwanic formations, represented by aeolian sandstones of Botucatu Formation and by basaltic rocks of the Serra Geral Formation.

These sedimentary facies characterize depositional systems of the lagoon/barrier type, associated to Upper Pleistocene and Holocene transgressive-regressive events.

RESUMO

A área mapeada neste trabalho está situada na porção nordeste do estado do Rio Grande do Sul, Brasil meridional, entre as latitudes de 29°15' e 29°45' sul e longitudes 49°30' e 50°15' oeste, incluindo os mapas planialtimétricos das Folhas de Torres, Tres Cachoeiras, Arroio Teixeira e Maquiné, na escala 1:50.000.

As principais unidades superficiais são os sedimentos marinhos rasos, eólicos, lagunares, deltáicos, fluviais e de leques aluviais, que são simbolizados respectivamente por

Qbc, Qbd, Qp, Qd, Qf, e Ql. Estes sedimentos estão sobrepostos as formações gonduânicas representadas pelos arenitos eólicos da Formação Botucatu e pelas rochas basálticas da Formação Serra Geral.

Estas fácies sedimentares caracterizam sistemas deposicionais do tipo laguna/barreira, associados aos eventos transgressivos-regressivos do Pleistoceno Superior e Holoceno.

INTRODUCTION

Since 1964, the researchers of the Center of Studies in Coastal and Marine Geology (CECO) had developed studies in southern Brazil continental shelf, in Rio Grande do Sul coastal plain including Patos Lagoon and in Antarctic Penin-sula, focalizing the main aspects of coastal and marine geology of these areas.

The knowledge and the paleogeographic evolution of the Coastal Province are the essential goals of the geological mapping, that is being executed through 640 km long of the present coastline of the state.

The area of this paper is situated on the southern Brazil, whose limits are in the north, the Santa Catarina State; in the south, Uruguai; in the west, Argentin and in the east, the Atlantic Ocean, according to the Figure 01.

The four quadrangles mapped, Torres, Tres Cachoeiras, Arroio Teixeira and Maquiné Sheets, are situated between the geographic coordinates of 29°15' - 29°45' south latitudes and 49°30' - 50°15' west longitudes.

The north limit of the area is 200 km from Porto Alegre, state capital, whose main access is the federal road BR-101.

METHODOLOGY

The methodology of the work follows six main stages: cadastration of the pre-existent basic data; preliminary photointerpretation; field work with sampling in surface outcrops; laboratory work comprehending mineralogic, granulometric, morphoscopic, geochronologic analyses; definitive photointerpretation and confection of the final geological 1:50.000 map that is reduced to 1:100.000 scale,

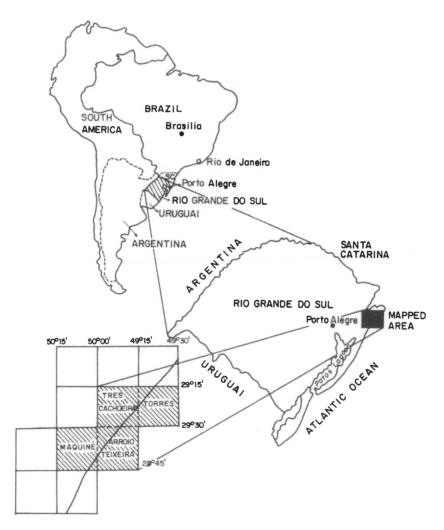

Figure 1. Localization of the mapped area.

congregating the four sheets in two groups, like is examined
in Photograph 1.

GEOMORPHOLOGY

The Figure 2 shows the four geomorphological provinces of the
Rio Grande do Sul state, according Carraro **et alii** (1974).
The Cambrian Cristaline Complex represented by the Sul-Rio-

Photograph 1. Geological map of the northeast coastal plain
of Rio Grande do Sul. Scale 1:100.000. According to: Horn F°
et alii (1984a and b).

Grandense Shield characterized the oldest unities. On this
shield settled by the Paleozoic and Mesozoic sedimentary and
volcanic sequences of Paraná Basin from the Devonian Period
onwards after the South American Platform stabilization. The
Serra Geral Plateau is the third province and consisted of
the tholeitic volcanism of the end of the Mesozoic Era. The
last and more recent geomorphological province is the Rio
Grande do Sul Coastal Province formed by two large geologic^
units: the Basement and the Pelotas Basin. The Coastal Plain
has 640 km long, a low relief and predominantly sandy. In this
plain is situated the Patos Lagoon, one of the largest lagoons
of the world.

Figure 2. Geomorphological provinces of Rio Grande do Sul
State. Compiled from: Carraro **et alii** (1974).

STRATIGRAPHY AND GEOLOGY

The studies that has been carried in the Coastal Plain of
Rio Grande do Sul shared the stratigraphy in two main subdivisions: the classical stratigraphy based on Delaney works
and the actual proposition based on Villwock works, that are
being developed by the researchers of the CECO.
  Delaney (1965) and another authors, such as Jost (1971a
and b), Villwock (1972), Soliani (1973), Godolphim (1976),
Jost and Soliani (1976), Ayala (1977) and Loss **et alii** (1982),
described unities of lithostratigraphic nature (Graxaim, Chuí,

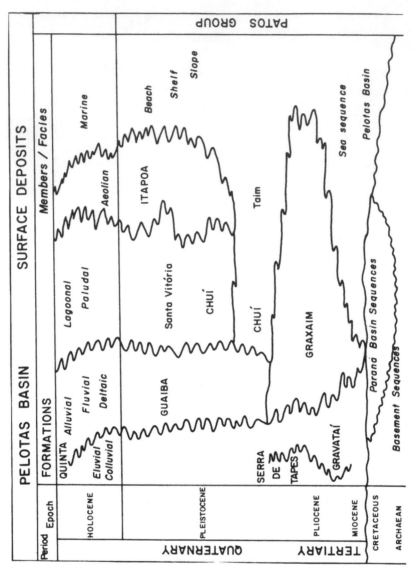

Figure 3. Stratigraphy of the surface deposits of the Pelotas Basin. According to: Villwock (1984).

Itapua, Guaíba Formations), whereas others of
chronostratigraphic nature like Quinta Formation, according
to Figure 3 scheme.

Villwock (1984) and another authors, such as Horn F° **et alii**
(1984a and b), Dehnhardt **et alii** (1984), Loss **et alii** (1984),
Tomazelli **et alii** (1984), Villwock **et alii** (1984), Loss **et**
alii (1985a and b), Villwock **et alii** (1985) and Villwock **et**
alii (1987), due to the difficulty in employing the subdivisions based on litho and chronostratigraphic criteria,
proposed the grouping of the units taking into account the
depositional systems with their respective sedimentary facies.

According to recent works, in Tomazelli **et alii** (1987) and
Villwock **et alii** (**op. cit.**), the Figure 4, shows the main
mapped facies that are: alluvial fan deposits (Ql), lagoonal
deposits (Qp and Qc), paludal and fluvial deposits (Qf), beach
deposits (Qbc), and transgressive marine deposits (Qm) that
do not crop out, separated in four, lagoon/barrier systems,
I-II-III-IV, respectively associated to Lower-Middle-Upper
Pleistocene and Holocene.

The geological map of the northeast coastal plain of the
state, 1:100.000 scale (Photograph 1) presents the main
unities of the mapped area: The Gondwanic Formations and the
Cenozoic deposits.

The highlands of the area which consisted the main source
area of the coastal plain are represented by two lithostratigraphic unities:

1. Botucatu Formation of the Jurassic and composed of aeolian
feldspathic cross-bedded sandstone, which in some places is
eroded by marine erosion in a highest sea level of Upper
Pleistocene.

2. Serra Geral Formation of the Cretaceous, represented by
tholeitic basalts that are the upper unities of the intra-
cratonic Paraná Basin, developed in southern Brazil. Torres
city, far 200 km away Porto Alegre, is the unique basalt
outcroping of South America in the present shoreline.

The Cenozoic deposits are subdivided in Pleistocene
sequences of the Lagoon/Barrier III System and facies of the
most recent depositional system developed during the Holocene.

The Pleistocenic deposits included the beach and intertidal
sediments, well sorted and constituted by fine to medium
quartzose clear sand, and the well rounded, well sorted red
sand dunes. During the Upper Pleistocene regression, the dunes
(Qbd3) developed over the marine terraces of flat top (Qbc3).

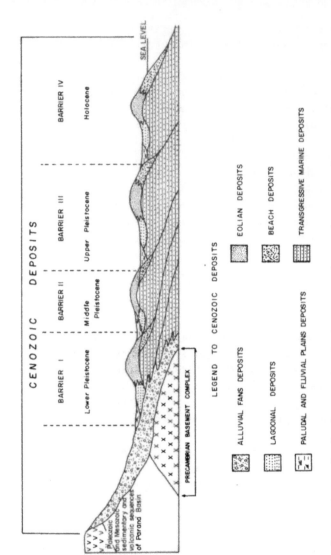

Figure 4. Geological sketch of the coastal province of Rio Grande do Sul.. According to: Villwock et alii (1987).

The Holocenic deposits of the Barrier IV grouped in previous works as belonging to chronostratigraphic unity called Quinta Formation by Godolphim (**op. cit.**), in the mapped area included:

- over the gondwanic formations and pleistocenic deposits, along of the present shoreline, are developed fine to medium quartzose sands of the aeolian dunes (Qbd4) and beach deposits presenting plano-parallel to cross bedding with intercalation of clear sand and dark sand rich in heavy minerals;
- regressive beach ridges deposits originated by the progradation of the Barrier IV (Qbc4);
- aeolian sand dunes string parallel to Itapeva and Quadros Lagoons (Qbd4);
- deltaic deposits (Qd4) produced by Maquiné and Tres Forquilhas rivers entering in lagoonal bodies, included gravel, pebbles of sandstone and basalts derived of Serra Geral Plateau;
- clayey and silty sands, poorly sorted of the lagoonal deposits (Qp4);
- massive reddish mudstones and sandstones of the alluvial fans associated to braided channels (Ql4);
- silt and clay tabular indifferenciated fluvial deposits (Qf4).

PALEOGEOGRAPHIC EVOLUTION

The Cenozoic deposits of the mapped area are correlationed to sea level changes of the Upper Pleistocene Transgression-regression event, individualizing the marine terraces and aeolian dunes of the Barrier III, evoluted mainly from the growing of sandy recurved spits.

The last transgression-regression event of the Holocene (5.500 years before present until today), was the responsible for the construction of marine beach ridge, aeolian, lagoonal, deltaic, fluvial and alluvial fan deposits correlationed to Lagoon/Barrier IV System.

MINERAL RESOURCES

The occurrence of mineral resources in the area and Coastal Plain of the state, are associated with the geological history of the coastal region and adjacent highlands.

The main resources included black sands in unexplorable beach deposits; peat deposits of lagoonal sediments used like corrective soil agent; sands for the glass industry and civil construction; clayey and silty sands for ceramic industry; flagstones and flintstones derived of sandstone Botucatu and basalts; water to the agriculture, pecuary and urban areas.

ACKNOWLEDGEMENTS

This work was presented in the Coastal Processes Session of the XII Quaternary International Congress, promotion of the International Union for Quaternary Research (INQUA).

My gratefulness to the Center of Studies in Coastal and Marine Geology, to the Institute of Geosciences and to the Federal University of Rio Grande do Sul for the opportunity; to the INQUA Comittee that accepted this paper and specially to the Interministerial Comission for the Resources of the Sea (CIRM) that supported my presence in Ottawa, Canadá.

REFERENCES

Ayala, L. 1977. **Contribuição ao Estudo da Formação Graxaim do Cenozóico da Planície Costeira do Rio Grande do Sul, Porto Alegre.** Curso de Pós-Graduação em Geociências UFRGS. 88pp. Dissertação Mestrado. Geociências. Porto Alegre.

Carraro, C.C.; Gamermann, N.; Eick, N.C.; Bortoluzzi, C.A.; Jost, H.; Pinto, J.F. 1974. **Mapa Geológico do Estado do Rio Grande do Sul - Escala 1:1.000.000.** Mapa nº 8, Instituto de Geociências, UFRGS, Porto Alegre - Brasil.

Dehnhardt, E.A.; Villwock, J.A.; Loss, E.L.; Hofmeister, T. 1984. **Mapa Geológico das Folhas de Gravataí e Santo Antônio da Patrulha.** In: CECO (ed.), Atlas Geológico da Provincia Costeira do Rio Grande do Sul. CECO/DGC; Instituto de Geociências, UFRGS. Porto Alegre.

Delaney, P.J.V. 1965. **Fisiografia e geologia da superfície da planície costeira do Rio Grande do Sul.** Publicação especial da Escola de Geología, UFRGS, 6:1-195. Porto Alegre.

Godolphim, M.F. 1976. **Geologia do Holoceno Costeiro do Município de Rio Grande - RS.** Curso de Pós-Graduação em Geociências, UFRGS, 146p., Fig. 1-37, Fot. 1-8, 1 mapa, tab. 1-9,

Dissertação Mestrado, Geociências. Porto Alegre.

Horn, F°, N.O.; Loss, E.L.; Tomazelli, L.J.; Villwock, J.A.; Dehnhardt, E.A.; Koppe, J.C. 1984a. **Mapa Geológico das Folhas Tres Cachoeiras e Torres.** In: CECO (ed.), Atlas Geológico da Província Costeira do Rio Grande do Sul. CECO/DGC; Instituto de Geociências, UFRGS. Porto Alegre.

Horn, F°, N.O.; Loss, E.L.; Tomazelli, L.J.; Villwock, J.A.; Dehnhardt, E.A.; Koppe, J.C.; Godolphim, M.F. 1984b. **Mapa Geológico das Folhas de Maquiné e Arroio Teixeira.** In: CECO (ed.), Atlas Geológico da Província Costeira do Rio Grande do Sul: CECO/DGC; Instituto de Geociências, UFRGS. Porto Alegre.

Jost, H. 1971a. **O Quaternário da Região Norte da Planície Costeira do Rio Grande do Sul - Brasil.** Curso de Pós-Graduação em Geociências, UFRGS. p81., Fig. 1-14, Fot. 1-16. Dissertação Mestrado. Geociências. Porto Alegre.

Jost, H. 1971b. **O Quaternário da Planície Costeira do Rio Grande do Sul. I. A regiao Norte.** In: Congresso Brasileiro de Geologia, 25. Anais... São Paulo, Sociedade Brasileira de Geologia. 1:53-62, Fig. 1-5. São Paulo.

Jost, H. & Soliani Jr., E. 1976. **Plano Integrado para o Desenvolvimento do Litoral Norte do Rio Grande do Sul: Mapeamento Geológico e Geomorfológico.** Secretaria de Coordenação e Planejamento - Fundação de Economia e Estatística do Governo do Estado do Rio Grande do Sul. p.121. Mapa Anexo.

Loss, E.L.; Dehnhardt, E.A.; Villwock, J.A. & Hofmeister, T. 1982. **Geologia da Bacia do Gravataí.** No prelo.

Loss, E.L.; Dehnhardt, E.A.; Villwock, J.A. & Hofmeister, T. 1984. **Mapa Geológico das Folhas Passo do Vigário e Lagoa do Capivari.** In: CECO (ed.), Atlas Geológico da Província Costeira do Rio Grande do Sul. CECO/DGC; Instituto de Geociências, UFRGS. Porto Alegre.

Loss, E.L.; Bachi, F.A.; Villwock, J.A.; Cunha, R.; Juchen, P.L. 1985a. **Mapa Geológico das Folhas de Itapuã e Desertas.** In: CECO (ed.), Atlas Geológico da Província Costeira do Rio Grande do Sul. CECO/DGC; Instituto de Geociências, UFRGS. Porto Alegre.

Loss, E.L.; Villwock, J.A.; Dehnhardt, E.A.; Tomazelli, L.J.; Godolphim, M.F.; Horn F°, N.O.; Bachi, F.A. 1985b. **Mapa Geológico das Folhas da Lagoa dos Gateados e Farol da Solidão.** In: CECO (ed.), Atlas Geológico da Província Costeira do Rio Grande do Sul. CECO/DGC; Instituto de Geociências, UFRGS, Porto Alegre.

Soliani Jr., E. 1973. **Geologia da região de Santa Vitória do Palmar, RS, e a posição estratigráfica dos fósseis mamíferos pleistocênicos.** Curso de Pós-Graduação em Geociências, UFRGS. p.88., pl.1-4. Dissertação Mestrado. Geociências. Porto Alegre.

Tomazelli, L.J.; Horn Fº, N.O.; Villwock, J.A.; Dehnhardt, E.A.; Loss, E.L.; Koppe, J.C. 1984. **Mapa Geológico das Folhas de Osório e Tramandaí.** In: CECO (ed.), Atlas Geológico da Província Costeira do Rio Grande do Sul. CECO/DGC; Instituto de Geociências, UFRGS - Porto Alegre.

Tomazelli, L.J.; Villwock, J.A.; Loss, E.L. 1987. **Roteiro Geológico da Planície Costeira do Rio Grande do Sul.** Publicação Especial nº 2. In: 1º Congresso da Associação Brasileira de Estudos do Quaternário, Julho 1987, Porto Alegre.

Villwock, J.A. 1972. **Contribuição à geologia do Holoceno da Província Costeira do Rio Grande do Sul.** p.113. Instituto de Geociências, UFRGS. Dissertação Mestrado. Geociências. Porto Alegre.

Villwock, J.A. 1984. **Geology of the Coastal Province of Rio Grande do Sul, Southern Brazil.** A Synthesis. Pesquisas, Instituto de Geociências, UFRGS, 16:5-49, Março, Porto Alegre.

Villwock, J.A.; Dehnhardt, E.A.; Loss, E.L.; Tomazelli, L.J.; Koppe, J.C. 1984. **Mapa Geológico das Folhas Rancho Velho e Cidreira.** In: CECO (ed.) Atlas Geológico da Província Costeira do Rio Grande do Sul. CECO/DGC; Instituto de Geociências, UFRGS, Porto Alegre.

Villwock, J.A.; Loss, E.L.; Dehnhardt, E.A.; Tomazelli, L.J.; Bachi, F.A.; Horn Fº, N.O. 1985. **Mapa Geológico das Folhas Ilha Grande e Balneário do Quintão.** In: CECO (ed.), Atlas Geológico da Província Costeira do Rio Grande do Sul. CECO/DGC; Instituto de Geociências, UFRGS. Porto Alegre.

Villwock, J.A.; Tomazelli, L.J.; Loss, E.L.; Dehnhardt, E.A.; Horn Fº, N.O.; Bachi, F.A.; Dehnhardt, B.A. 1987. **Geology of the Rio Grande do Sul Coastal Province.** In: Quaternary of South America and Antarctic Peninsula - Vol. 4, Selected Papers of the International Symposium on sea-level changes and Quaternary shorelines, 7-14 July 1986. p.344. São Paulo.

Milton Keynes UK
Ingram Content Group UK Ltd.
UKHW040711141024
449569UK00005B/106